Creative Industries and Urban Development

The concept of creative industries has developed considerable academic and policy momentum in the 21st century. There has been a connection identified between the rise of creative industries and the urbanisation of the world's population, particularly in relation to the significance of cities as sites of cultural production and consumption. Much of the work on creative industries and cities, however, has drawn upon 'imagined geographies' about the relationship between creativity and place. This collection draws together contributions that critically appraise recent urban cultural policy discourses, as well as reflecting on the role of culture and creative industries in the future development of cities.

This book is based on a special issue of *The Information Society: An International Journal*.

Terry Flew is Professor of Media and Communications at the Queensland University of Technology, Australia. He is the author of *The Creative Industries: Culture and Policy* (2012), *New Media: An Introduction* (2008) and *Understanding Global Media* (2007), and is a leading international figure in creative industries research. From 2008–2010 he led an Australian Research Council project into creative industries and suburban development. During 2011–2012, he headed a review of media classification for the Australian Federal government.

Creative Industries and Urban Development

Creative Cities in the 21st Century

Edited by
Terry Flew

Routledge
Taylor & Francis Group

LONDON AND NEW YORK

First published 2013
by Routledge
2 Park Square, Milton Park, Abingdon, Oxfordshire OX14 4RN

Simultaneously published in the USA and Canada
by Routledge
711 Third Avenue, New York, NY 10017

First issued in paperback 2014

Routledge is an imprint of the Taylor & Francis Group, an informa business

© 2013 Taylor & Francis

Chapters 2, 3, 5 and 7 of this book are a reproduction of *The Information Society: An International Journal*, volume 26, issue 2. The Publisher requests to those authors who may be citing these chapters to state, also, the bibliographical details of the special issue on which the book was based.

British Library Cataloguing in Publication Data
A catalogue record for this book is available from the British Library

ISBN 978-0-415-51680-8 (hbk)
ISBN 978-1-138-84177-2 (pbk)

Typeset in Times New Roman
by Saxon Graphics Ltd, Derby

Publisher's Note
The publisher would like to make readers aware that some chapters in this book may be referred to as articles as they are identical to the articles published in the special issue. The publisher accepts responsibility for any inconsistencies that may have arisen in the course of preparing this volume for print.

Contents

Notes on Contributors

Chris Brennan-Horley is GIS Project Manager for the Australian Centre for Cultural Environmental Research (AUSCCER) at the University of Wollongong, Australia. He maintains a research interest in applying geographic information systems to cultural research, particularly around issues relating to creative industries and cities.

Christy Collis is a Senior Lecturer in Media and Communication, and Head of Entertainment Industries in the Creative Industries Faculty, Queensland University of Technology, Brisbane, Australia. Christy is a cultural geographer. With Professors Terry Flew and Phil Graham, and Mark Gibson and Emma Felton, she recently completed an Australian Research Council project titled Creative Suburbia, which examined creative industries work and workers in Australian outer suburbs.

Stuart Cunningham is Distinguished Professor of Media and Communications, Queensland University of Technology, Australia, and Director of the Australian Research Council Centre of Excellence for Creative Industries and Innovation. His most recent books are *Hidden Innovation: Policy, Industry and the Creative Sector* (2012), *The Media and Communications in Australia* 3rd edition (with Graeme Turner, 2010) and *In the Vernacular: A Generation of Australian Culture and Controversy* (2008).

Emma Felton is Co-ordinator of the Widening Participation Program and Research Fellow in the Creative Industries Faculty at Queensland University of Technology, Australia. She has a PhD in Cultural Geography and researches and writes on urbanism from socio-cultural perspectives. Emma has also worked in e-learning leadership in the university sector, and has received grants for learning and teaching and in the field of communication technologies.

Terry Flew is Professor of Media and Communications at the Queensland University of Technology, Australia. He is the author of *The Creative Industries: Culture and Policy* (2012), *New Media: An Introduction* (2008) and *Understanding Global Media* (2007), and is a leading international figure in creative industries research. From 2008–2010 he led an Australian Research Council project into creative industries and suburban development. During 2011–2012, he headed a review of media classification for the Australian Federal government.

Chris Gibson is Professor in Human Geography and Deputy Director of the Australian Centre for Cultural Environmental Research (AUSCCER) at the University of Wollongong, Australia. His most recent books include *Creativity in Peripheral Places: Redefining the Creative Industries* (Routledge, 2012) and *Music Festivals and Regional Development in Australia* (Ashgate, 2012).

Phil Graham is Professor of Music and Sound in the Creative Industries Faculty, Queensland University of Technology, Australia. Prior to entering academia, he spent 20 years as a professional musician working as a performer, composer, and producer. He is Editor of *Critical Discourse Studies* and an International Advisory Board Member of multiple journals in discourse, media, cultural, and business studies.

Xin Gu's research focuses on creative industries and creative entrepreneurship in the city. She has done extensive work on culture led urban regeneration in Manchester, Tilburg, Helsinki and Brussels. She was Senior Research Associate in an Australian Research Council Linkage project on Designing Creative Clusters in China and Australia (2010–2012).

Angela Lin Huang is a Researcher at Beijing Research Centre for Science, Beijing Academy of Science and Technology. She worked on the editing of Chinese Creative Industries Report from 2006 to 2010. She completed her PhD at Creative Industries Faculty, Queensland University of Technology in 2012, studying Beijing's comparative advantage as an emerging media capital.

She has also participated in dozens of national research projects and Beijing municipal research projects regarding creative industries. Her research interest covers creative industries, media studies, industrial policy and innovation strategy.

Susan Luckman is Portfolio Leader: Research and a Senior Lecturer in the School of Communication, International Studies and Languages at the University of South Australia who teaches and researches in the fields of media and cultural studies. She is author of the forthcoming book *Place and Cultural Work: The Politics and Poetics of Rural, Regional and Remote Creativity* (Palgrave Macmillan); co-edited the anthology on creative music cultures and the global economy (*Sonic Synergies*, Ashgate 2008); and is the author of numerous book chapters, peer-reviewed journal articles and government reports on creative cultures and industries.

John Montgomery is an urban economist and town planner who has specialised over many years in the creative industries, economic development and urban regeneration. Much of his experience is drawn from the United Kingdom and Ireland, dating from the mid 1980s, and with his London-based firm Urban Cultures Ltd since 1991. He prepared the first creative industries economic development strategy in the UK, for Manchester and the North West, in 1988–89, and went on to produce the first cross-London study of, and strategy for, the creative industries, in 1992–93 and published in 1994. This set the definitions and methodologies for the creative industries as a sector of the economy, since taken up officially by the UK government. Other more localised studies followed, for West London in 1995, Birmingham in 1998 and Sheffield. He is an Adjunct Professor with the Queensland University of Technology, Australia. His research interests most recently lie in the field of evolutionary economics and the growth of cities over long waves of economic development.

Justin O'Connor is Chair of Communications and Cultural Economy in the School of English, Performance and Communication Studies, Monash University. He is also visiting Chair in the Department of Humanities at Shanghai Jiao Tong University, China. Until 2012 he was Professor in the Creative Industries Faculty, Queensland University of Technology, and between 2006–2008 Chair in Cultural Industries at the University of Leeds. Before that he was Director of the Manchester Institute of Popular Culture for ten years. He is currently one of 20 international experts appointed under the UNESCO/EU Technical Assistance Programme in support of the 2005 Convention on the Protection and Promotion of the Diversity of Cultural Expressions.

Julie Willoughby-Smith graduated from the University of South Australia with a Bachelor of Arts Degree majoring in Communication and Media Management. She worked as a Research Assistant for the School of Communication, University of South Australia on the Australian Research Council project, Creative Tropical City: Mapping Darwin's Creative Industries and was involved with fieldwork, ethnographic interviews and the collection of data. Her research interests include the socio-cultural dimensions of globalisation, the creative industries, new media communications, mobilities, travel and tourism studies.

Introduction: Creative Industries and Cities

Terry Flew

Creative Industries Faculty, Queensland University of Technology, Brisbane, Queensland, Australia

CREATIVE INDUSTRIES AND CITIES: A DOMINANT *MOTIF* OF THE 2000S

The 2000s were marked by a resurgence of interest in creativity and cities. If the rapid global proliferation of the Internet and digital media technologies in the 1990s had set off enthusiasm for a post-industrial 'new economy', where the significance of location would be in decline, the 2000s saw an energetic search by artists, entrepreneurs, investors, policy-makers, journalists and many others to uncover the well-springs of creativity and its relationship to place (Flew 2012a). This chapter begins with a discussion of the discourses or 'scripts' that have emerged to try and conceptualise the relationship between creativity and cities, notably theories of creative clusters, creative cities and creative class theories. Such work can be seen as representing a growth in the field of cultural economic geography although – as is noted in the chapter – it possesses some significant gaps. Among the issues that are drawn out in this book, and discussed in this chapter, are: the need to move beyond 'imagined geographies' of creative inner cities and come to terms with empirical evidence that suggests significant concentrations of the creative workforce in suburbs and regional cities; the relevance of urban cultural policy as a variable in the rise of cities as creative hubs or, in a different model, media capitals; and the challenges of bringing together cultural research with economic discourses in ways that get beyond caricatured representations of the 'other', as found, for instance, in some of the most influential framings of the concept of neo-liberalism.

Creativity was seen as the foundation of innovation, and innovation was seen as the new primary driver of economic growth. If an exemplary 'new economy' business of the 1990s was Microsoft, attention had turned by the 2000s to the social media businesses such as Google, Twitter and Facebook, which grew not by making established products, services and processes better, but by developing entirely new ways of doing things, or completely new things to do, like participating in online social networks rather than reading newspapers, or viewing amateur videos online rather than watching television.

At the core of all of this was human creativity, described by creative economy guru Richard Florida as the most elusive resource:

> Creativity has come to be the most highly prized commodity in our economy – and yet it is not a 'commodity'. Creativity comes from people. And while people can be hired and fired, their creative capacity cannot be bought or sold, or turned on or off at will … Creativity must be motivated and nurtured in a multitude of ways, by employers, by people themselves and by the communities where they locate. Small wonder that we find the creative ethos bleeding out from the sphere of work to infuse every corner of our lives (Florida 2002: 5).

Renewed interest in creativity has coincided with what Allen Scott (2008) refers to as the *resurgence of cities*. While much talk in the 1970s and 1980s was about the crisis of cities, faced with the shift of manufacturing to lower-wage economies and the decline of inner cities, and the 1990s saw prophecies that the Internet was heralding the 'death of distance' (Cairncross 1998), what has become apparent is that globalisation, the rise of digital media networks and industries, and the need to develop post-industrial urban development strategies have all contributed to cities becoming 'motors of the global economy' (Scott *et al.* 2001). Over the decades the trends in global population distribution were also moving in this direction. In 1950, less than 30 per cent of the world's population lived in cities. By 2007, the percentage of the world's population living in cities exceeded those living in rural areas, for the first time in human history (Worldwatch Institute 2007).

What was notable was the symbiotic relationship that that been seen to exist between creative industries and cities. Spatial agglomeration, or clustering, was seen enhancing innovation and flexibility by promoting information flows, networks of interaction and relational ties among a diverse but spatially proximate range of participants and institutions, specially in industries characterised by high levels of uncertainty, instability, and complexity (Scott *et al.* 2001). The British economist Alfred Marshall had noted more than a century ago the positive externalities that can arise from a clustering of related firms and industries in a particular location

(Marshall 1990 [1890]). But the case of creative industries differed in that they are driven by the externalities that arise not only from specialisation in particular industries and occupations, but also from the positive externalities that arise from the diversity of cities themselves (Lorenzen and Frederiksen 2008). With their diversity of industries, forms, workforce and skills, as well as cultural diversity, cities can be centers for co-ordination between diverse knowledge bases, as the geographical proximity of people promotes knowledge flows, the spread of ideas, and new forms of entrepreneurship. As much work in the creative industries is project-based, contractual and time-dependent (Caves 2000), advantages exist for small-to-medium enterprises (SMEs) and workers in the creative sectors through clustering in locations where work emerges on a regular basis. This in turn means that cities act as 'talent magnets' for skilled people from other parts of the country and the world. There is also a positive correlation between cities being the centers of financial and professional services and the arts and entertainment industries, both because cultural services are most easily accessed from these centers, and because they typically have high average levels of consumption of cultural goods and services.

Cities tend to be the centers of what Landry (2000) termed the 'hard infrastructure' of creative industries, as they are typically where the head offices of the major industry players are located (especially in media-related sectors) and where governments have typically invested heavily in the cultural infrastructure of cities, with their extensive network of galleries, museums, libraries, universities etc. This can in turn act as a catalyst for the formation of 'soft infrastructure', or the *relational assets* associated with economically successful networks, such as trust, reciprocity, exchange of tacit knowledge, and propensity to share and pool economic risk (Amin 2003). They have also been, historically, the centers of culture. Hall (1998) observed that because the city 'continues to attract the talented and the ambitious … it remains a unique crucible of creativity' and, through his historical account of great cities, he argued that

> while no one kind of city, or any one size of city, has a monopoly on creativity or the good life … the biggest and most cosmopolitan cities, for all their evident disadvantages and obvious problems, have throughout history been the places that ignited the sacred flame of human intelligence and the human imagination. (Hall 1998: 7)

THE CLUSTER SCRIPT

The first framework that was used to understand the relationship between creative industries and cities was that of *clusters*. As noted above, an interest in clusters among economists and geographers can be dated back to the work of Alfred Marshall on industrial districts in the late nineteenth century, although the concept has had an uneven history, perhaps because – at least for economists – it challenges the equilibrium modeling of neo-classical theories by pointing to the prospect of 'winner takes all' outcomes. In the 1990s, the cluster concept experienced a resurgence through the work of business management theorist Michael Porter of the Harvard Business School. In extending his *competitive advantage* model from firms to nations, Porter argued that the dynamic and sustainable sources of competitive advantage derived less from lower costs and production efficiencies than it did from elements of a place that promote productivity growth and innovation over time. In particular, and following Marshall, Porter was interested in the *spillover* benefits that can emerge from being in particular locations, which have related and supporting industries.

Porter (1998, 2000) argued that clusters are able to provide three sources of competitive advantage to the firms that are a part of them:

1. *productivity gains*, deriving from access to specialist inputs and skilled labor, access to specialised information and industry knowledge, the development of complementary relationships among firms and industries, and the role played by universities and training institutions in enabling knowledge transfer
2. *innovation opportunities*, derived from proximity to buyers and suppliers, sustained interaction with others in the industry, and pressures to innovate in circumstances where cost factors facing competitors in other locations are broadly similar
3. *new business formation*, arising from access to information about opportunities, better access to the resources which are required by business start-ups (e.g. venture capitalists, skilled workforce), and reduced barriers to exit from existing businesses, since takeovers and mergers are more readily facilitated due to geographical proximity between large and small firms in the industry.

The notion of creative clusters has lent itself well to strategies of culture-led urban regeneration that have been a feature of post-industrial cities in Europe in particular, stimulated by the European Union through initiatives such as the European Capital of Culture program (Mommaas 2009; Palmer and Richards 2011). Cluster development had a strong appeal to urban policy-makers, and this was consistent with sub-national levels of government increasingly becoming engaged in cultural policy in an era of economic globalisation (Schuster 2002). As the creative cities literature has often been characterised as being 'heavily reliant on proxies but light on theory or hard evidence' (Evans 2009: 1005), cluster theory generated no shortage of international exemplars, such as the Hollywood film and TV cluster, the high-technology cluster of 'Silicon Valley', the design and advertising clusters of London, and the fashion districts of Paris and Milan. It also seemed to generate a strong momentum in countries where a collectivist ethos has long been cultivated by governments, such as Singapore and (especially) China, as the ways in which 'in the cluster literature, social networks, tacit knowledge and trust relationships are valorised' (Kong 2009: 70) are consistent with both state ideologies and Confucian ideas that promote working together around shared problems and common goals, in contrast to Western liberal individualism (Keane 2009).

The motivations behind creative cluster development were mixed, ranging from city branding strategies to building new forms of cultural infrastructure, promoting cultural diversity to redeveloping derelict industrial-era sites such as warehouses and power stations. Such sites commonly had a 'post-industrial' mix of activities including residential apartments, arts centers and business incubators (Mommaas 2009). Given the eclectic range of motivations that lay behind creative cluster developments, it was not surprising that the scorecard for these new 'creative' urban cultural policies was also mixed (Bassett *et al.* 2005). Some of the benefits have included: a greater role for culture in urban development strategies; a broader and more inclusive understanding of culture than simply the 'high arts'; greater recognition of lifestyle factors and consumption activities in urban planning; and the development of new cultural infrastructures that have renovated the image of cities and acted as attractors of tourism and private investment. Problems with creative cluster policies have included: a blurring of the distinctiveness of arts and culture into entertainment, leisure and service industries; possibly contradictory policy agendas between economic development and social inclusion; instances of 'capture' of the urban renewal agenda by private real estate interests; and the possibility that the drive to develop distinctive creative clusters has the paradoxical effect of promoting greater urban homogeneity. The latter is what Kate Oakley referred to as a 'cookie-cutter' approach to developing the creative industries, characterized by 'a university, some incubators, and a "creative hub", with or without a café, galleries and fancy shops' (Oakley 2004: 73).

Some of the problems arose squarely from the cluster concept itself, and the ways in which the concept has developed in such a loose and all-inclusive manner that, as Martin and Sunley have observed, 'it is impossible to support or reject clusters definitively with empirical evidence, as there are so many ambiguities, identification

problems, exceptions and extraneous factors' (Martin and Sunley 2003: 31). One problem is the propensity to conflate geographical and industrial definitions of a cluster, so that there is a failure to distinguish between clusters where a number of firms in the same industry have co-located (*horizontal clusters*), such as the successful wine industries of Northern California in the United States and the Barossa Valley in Australia, and those where a value chain of buyers and suppliers has emerged (*vertical clusters*), such as the ICT/electronics hub of Silicon Valley. While both types of cluster enable knowledge transfer to occur, they do so in quite different ways, and this is blurred by the concept of creative clusters being associated with a highly diverse and in many ways disconnected set of 'creative industries'.

Moreover, agglomeration is not in itself evidence of clustering in the manner that Porter refers to it. Gordon and McCann (2001) distinguish between what they refer to as simple agglomeration, where co-location in particular areas reduces overall costs (e.g. transport and catering businesses clustering around an airport), and those where it is social networks and embedded ties that are critical to locational decisions: it would only be in the latter case where clustering would be strongly connected to innovation through knowledge flows. Finally, it remains unclear whether particular cities develop successful creative clusters because they are global cities, as with London, New York and Paris, and whether what Stevenson (2002) terms the 'civic gold rush' to build creative clusters in the hope of attracting major creative industries firms away from these global centers will run into difficulties in the face of powerful forces promoting agglomeration and sustained competitive advantage in the established urban spaces.

THE CREATIVITY SCRIPT

If clustering was one of the more common explanations for the tendency towards the agglomeration of creative industries in urban centers, the other was linked to the concept of creativity. As noted above, the work of Richard Florida (2002, 2008) was central to this, as it inverted the standard script on urban economic development that pointed to the need for subsidies and tax breaks to entice large employers. Florida instead argued that the growth potential of cities derived from their attractiveness to creative people or what he called the *creative class*. This argument generated a strong groundswell of interest among urban planners and policy-makers (Peck 2005), and resonated with the push to develop creative cities associated with consultancy groups such as Comedia in Britain (Landry 2000), the European Capital of Culture initiatives and the redevelopment of cities such as Dublin and Barcelona. Florida's work owed a considerable debt to earlier theorists of the city, such as Jane Jacobs, who saw creativity in cities arising out of the mix of proximity, diversity and sociality that marked their populations, as well as the importance of 'third places' between home and work as sites that sparked new social networks and the formation of new forms of community (Florida 2008). Such arguments paralleled the emphasis in Charles Landry's work on the importance of *creative milieux* and 'soft infrastructure' in the creative industries, with the latter defined as 'the system of associative structures and social networks, connections and human interactions, that underpins and encourages the flow of ideas between individuals and institutions' (Landry 2000: 133).

Florida's analysis of the role of the creative class and the rise of creative cities has been widely debated and hotly disputed. For those who are skeptical of the wider claims being made about creativity, there is the significant problem of the lack of any clear and widely accepted measures of creativity that go beyond the anecdotal and impressionistic (Galloway and Dunlop 2008). Alternatively, there is the danger that Florida has cast his net far too widely in defining a 'creative class', and that it too easily becomes a proxy for most people with a higher degree. Pratt (2008) has critiqued the focus of this approach upon the consumption choices of the urban middle classes, arguing instead that the focus needs to be on cultural production and questions of how and why it locates in particular geographical areas. Peck (2005) has critiqued Florida's account as a 'fast policy' script for urban policy that shows little concern for those not in occupations or life situations that give them spatial mobility, and critiques what he terms *Homo Creativus* as 'an atomized subject, apparently, with a preference for intense but shallow and noncommittal relationships, mostly played out in the sphere of consumption and on the street' (Peck 2005: 746). This concern that 'the most creative places tend also to exhibit the most extensive forms of socio-economic inequality' (Peck 2005: 746) is echoed by Storper and Scott (2009), who argue that what they term *amenities-based* models of urban growth (those which focus on the supply of cultural goods and services and how they influence urban migration patterns) not only wrongly assume that all cities can follow a similar developmental trajectory, but that:

> The emerging new economy in major cities has been associated with a deepening divide between a privileged upper stratum of professional, managerial, scientific, technical and other highly qualified workers on the one side, and a mass of low-wage workers – often immigrant and undocumented – on the other side. The latter are not simply a minor side effect of the new economy or an accidental adjunct to the creative class. Rather, high-wage and low-wage workers are strongly complementary to one another in the new economy (Storper and Scott 2009).

CULTURAL ECONOMIC GEOGRAPHY

There is a need to take discussions about creative industries and their impacts on urban development beyond the clustering and creativity scripts. What is required, and which the essays in this collection seek to advance, are methodologies that move beyond metaphors and archetypes towards grounded empirical work, combined with interdisciplinary frameworks that are flexible and adaptive to trends in the cultural and economic spheres. One such reference point can be found with the framework of *cultural economic geography* which can enable enabling links to be made between insights from media, communications and cultural studies with concepts derived from areas of the social sciences such as institutional economics and public policy studies.

In their overview of the rise of cultural economic geography, James *et al.* (2008) distinguish five distinct but related factors associated with this emergent hybrid discipline. First, they observe that the 'Marxist turn' in economic geography in the 1970s and 1980s, which had sought to map spatial relations under capitalism to develop a historical materialist geography (e.g. Harvey 1982), was being challenged in the 1990s by the 'cultural turn' associated with post-structuralism, which sought to challenge some implicit hierarchies of thought in the dominant forms of critical geography (Gibson-Graham 2000). In particular, the post-structuralists questioned the discursive construction of 'the economy' in political economy, and what it prioritised and what it downplayed. For example, taking the category of 'labor', is paid wage-labor more significant than domestic labor, or is the fact of laboring more 'real' than the ways in which it is understood and approached in labor market theories, management discourses or policy-related definitions of work that impact upon welfare policies? Second, attention was drawn to the particular ways in which culture and economy interlock, such as the relationship between markets and production as spatially grounded economic practices and the lived experience of people within such economic spaces (c.f. du Gay and Pryke 2002). Third, the cultural constitution of economic practice, and the awareness that 'cultural' factors can mark significant sources of regional differentiation, local entrepreneurship and competitive advantage in globalised economies, as seen in the debates surrounding clusters and learning regions (Cooke and Lazzeretti 2008), as well as considerations of the cultural geography of economic production (Gertler 2003). Fourth, the rise of *actor-network theory* has been significant in focusing renewed attention upon the performative dimensions of 'soft capitalism' and the ways in which it is engaged in new business management practices (Latour 2007; Thrift 1999, 2005). Finally, there has been growing interest among academics and policy-makers in the creative industries has intensified interest in geographical location decisions surrounding these industries, and the relationship between the attributes of the industries [e.g. their propensity for project work, networking, unpredictability of demand, need for continuous novelty and innovation (Caves 2000)], and the attributes of urban environments in which they are primarily located.

A growing interest in spatial perspectives on economy, society and culture can be seen as consistent with what Ed Soja (2010) has referred to as the *spatial turn* in social theory. Soja has proposed that, 'whatever your interests may be, they can be significantly advanced by adopting a critical spatial perspective (Soja 2010: 2). This goes beyond the insistence that 'space matters', and that a spatial dimension needs to be incorporated into social theory. Rather, by drawing attention to the dynamic properties of socially constructed space itself, or what Soja terms 'the powerful forces that arise from socially produced spaces such as urban agglomerations and cohesive regional economies', it is proposed that 'what can be called the stimulus of socio-spatial agglomeration is today being assertively described as the *primary* cause of economic development, technological innovation, and cultural creativity' (Soja 2010: 14). This has implications for thinking about culture and economy more generally, which will be returned to below.

BEYOND IMAGINED GEOGRAPHIES: SUBURBS AND REGIONAL CITIES

A major problem with creative cities discourse, in noting that creative industries are frequently aggregated in major urban centers, is that it is unable to address the question of causality. In other words, do creative industries cluster in global cities because they are global centers of commercial activity, or do particular cities become centers of global commercial activity because of their cultural features and the creative attributes of their populations? It is apparent, for instance, that the arts and entertainment thrive in cities such as New York and London in part because they are also centers for financial and professional services. If this is the case, then how transferable are the experiences of these cities that are held up as exemplars (e.g. Landry 2005; Currid 2007) of the creative cities model?

One of the most sustained critiques of Richard Florida's work is that it presumes that the 'creative class' actively seeks out inner city living, in search of cultural amenities and 'buzz'. There has been a strong counter-argument, associated particularly with Joel Kotkin (2007), that the renewal of inner cities as residential area is a less significant cultural force than the 'new suburbanisation', or the demand for affordable housing of a reasonably large size, that is driven as much by the pull factor of suburban amenity as it is by the push factor of affordability. Such arguments gain force from the proposition that the most economically dynamic U.S. cities were not, at least until 2008, the more 'hip' and spatially dense ones such as San Francisco, Boston or Seattle, but more suburbanised ones such as Los Angeles, Dallas-Fort Worth and Phoenix. There is the broader empirical question as to whether differential growth rates between cities actually reflect different levels of human capital rather than creative capital i.e. cities with more highly educated populations are more prosperous than those with lower levels of education, with cultural factors playing only a minor role, as urbanists such as Edward Glaeser have long argued (e.g. Glaeser 2011).

In this account, the neglect of suburbs is as much a reflection of an imagined geography of exciting and diverse cities versus boring and homogeneous suburbs – one with deep roots in the history of suburbia (Clapson 2003) – as it is reflective of where creative workforce actually locates. This dualism has a tenacious history in cultural theory, where the city is viewed as dynamic, edgy, diverse and spontaneous, and hence conducive to creativity, whereas suburbs are viewed as static, dull, homogeneous and overly planned, whether by city authorities primarily concerned with a 'safe' environment, or by the developers of master-planned estates seeking to realise land values over a medium-term time horizon (Johnson 2012). While Florida's creative cities thesis may be seen as a 'boosterish' promotion of urbanity, such claims also pervade radical critiques of the 'creative city'. To take one prominent example, the radical geographer David Harvey's contemporary manifesto for a 'right to the city' (Harvey 2008), suburbs basically feature as an inferior form of urban living, characterised by rampant real estate speculation, political conservatism, 'big box' shopping complexes, desperate housewives and 'pacification by cappuccino' (Harvey 2008: 32). Their only redeeming feature, for Harvey, is that their 'soulless quality' makes them a fertile breeding ground for future radicals from middle-class homes, once they leave the suburbs and become politicised and more exposed to marginalised urban communities (Flew 2011).

The chapter by Christy Collis, Emma Felton and Phil Graham reports on a study being undertaken into creative workforce in selected Australian suburbs, which suggests that the common association of creative workers with inner city areas, and the urban cultural policy implications arising from such assumptions, need to be significantly rethought. They draw upon the 'new suburbanism' thesis, observing that much of what is understood as urban growth is in fact suburban growth: they quote Kotkin's (2005) observation that more than 90 per cent of metropolitan growth in the United States since 1950 has been in the suburbs, as well as Australian evidence that both population growth and jobs growth is increasingly located in the middle and outer suburbs of Australia's major cities, particularly in master-planned estates on the urban fringe.

Importantly, they present evidence demonstrating that this is in fact where the creative workforce is increasingly locating itself. This presents the question of whether they are forced out to the suburbs by high inner-urban rents and house prices, or drawn to the suburbs by aspects of the locality and lifestyle. Based upon extensive interviews conducted with creative workers in selected Melbourne and Brisbane suburbs, their finding is that it is overwhelmingly the latter factor that drives these location decisions. While inner cities remain at the centre of much activity in the diverse range of sectors in which they work – from architecture and design to the arts and music – the suburbs provide a range of affordances that act as stimuli to creativity, ranging from the relatively mundane, such as less time spent travelling, to the 'serenity' of such locations itself being more conducive to creative practice. In contrast to the claims about suburban homogeneity, they report a greater ability to 'be yourself' and not have to conform to peer pressures as being one of the more positive aspects of being located in the suburbs rather than in the inner city.

What renders this question more than simply hypothetical is that urban policy 'scripts' are derived from the experiences of such cities as guides for urban policy practice in very different cities, even though they will never acquire the global city status of the largest urban agglomerations. The challenge, as Chris Gibson has observed, is to get beyond the archetypes and stereotypes where:

> Researchers have looked for creativity in fairly obvious places (big cities, cities making overt attempts to reinvent themselves through culture, creativity and cosmopolitanism); have found it there; and have theorized about cities, creative industries and urban transformations as if their subsequent models or logic were universally relevant everywhere (Gibson 2010: 3).

The work on the tropical city of Darwin in northern Australia discussed by Chris Brennan-Horley, Susan Luckman, Chris Gibson and Julie Willoughby-Smith in this collection considers the implications of looking at creative industries and the creative workforce in a smaller and very different type of city. Darwin is nether a global commercial hub nor a city grappling with the implications of deindustrialisation, but which has a very distinctive set of historical, geographical and demographic features which nonetheless can act as catalysts to creative industries development, albeit in very different ways to the dominant interpretations.

What the work of Brennan-Horley *et al.* also brings out is the manner in which creativity can be grounded in place (cf. Drake 2003; Chapain and Communian 2010). Rather than starting from a series of policy indices of where the creative industries are located, such as the Location Quotient used by economic geographers (Collis *et al.* 2012), or from the Florida-inspired 'recipes' for creative place development (arts hubs, night life, cycle paths etc.), they deploy innovative methodologies such as the use of Geographical Information Systems (GIS) and 'mental mapping' to visualise how a sense of belonging, connection and place operates among Darwin's creative workers themselves, in order to capture the 'un-sexy' realities of sustainable creative practice in a particular location.

Both Collis *et al.* and Brennan-Horley *et al.*'s chapters draw attention to the extent to which uncovering creative activity outside of the core metropolitan centres may require methodological innovation in the analysis of creative industries and cities. In particular, both draw out the extent to which more qualitative social research methods are required, that enable practitioners themselves to speak for themselves, in order to get beyond imagined geographies of creativity and culture.

URBAN CULTURAL POLICY AS A RELEVANT VARIABLE

The existence of urban policy scripts such as those surrounding clusters and creativity that we have discussed draws attention to the role played by what Pratt (2009) terms *policy transfer*, and the role played by consultants and policy entrepreneurs in enabling ideas and policy prescriptions to travel from one context to another. For example, European Union funding associated with European Capital of Culture initiatives acted as a catalyst for the redevelopment of cities as diverse as Dublin, Glasgow, Madrid, Liverpool and Istanbul as centres of culture and creativity (Palmer and Richards 2011). Similarly, the aspirations for economic ascendancy in the Asia-Pacific region promoted creative industries policy 'scripts' in cities such as Singapore, Seoul, Shanghai and Auckland (Kong *et al.* 2006).

Cunningham and Flew observe in their chapter on the evolution of creative industries discourse how the original conception of creative industries as developed by the Department of Culture, Media and Sport in the United Kingdom in the late 1990s has subsequently been taken up in other parts of the world, and the significance of local inflections upon that original 'master discourse'. Drawing upon a range of international policy case studies, this essay questions the degree to which international uptake can simply be understood as simply entailing the international transfer of British 'New Labour' policy discourses, as argued by Garnham (2005) and Ross (2007). Instead, it is argued that creative industries, and associated concepts such as creative economy, have been tied up with a wider rethinking of the enabling factors for innovation in post-industrial economies, which engage the arts and humanities as well as the sciences and technology sectors. While proposing economic rationales for cultural investments is frequently derided as neo-liberalism by its critics, this term has become too all encompassing to be of analytical value. In particular, it wrongly presupposes a dichotomy between the public and private sectors, whereas creative industries policies by their nature have necessitated thinking about the commercial realm and the public sector in tandem as drivers of cultural innovation.

John Montgomery provides a spirited historical account of the significance of London as a creative city over several centuries. He locates the dynamism of London as both a commercial centre and a centre for the arts, media and culture in its ability to benefit from long waves of capitalist development. The concept of long waves, of 50-year cycles of capitalist boom and downturn, was pioneered by the Russian economist Nikolai Kondratiev and the famous Austrian economist Joseph Schumpeter; in recent years, it has been used by authors such as Manuel Castells (1996) and Carlota Perez (2010) to identify why clusters of technological innovation occur at particular times and how the diffusion of these core technologies sets off wider processes of socio-economic transformation. In contrast to more pessimistic accounts of London's future, Montgomery argues that the city is well placed to be at the centre of the 'Fifth Kondratiev Wave', centred around digital technologies and creative industries, due to the industry clusters and inter-firm networking relationships that have emerged in this dynamic and ever-changing global city.

The question of how best to develop creative industries policies invokes familiar tensions between top-down policy discourses developed at national government level, and strategies applied at the local level of the city-region. In the case of creative industries policies, this is overlaid with the related tensions as to whether the primary focus of policy is upon the cultural development of a city or its economic development, and implications of such decisions for who is at the centre of strategic conversations. In their comprehensive account of the rise and fall of the Creative Industries Development Service (CIDS) in Manchester in the UK, Justin O'Connor and Xin Gu capture the extent to which, while these can sometimes converge, they can and often do significantly diverge. In particular, they argue that the 'local knowledge' that CIDS was able to draw upon in brokering a relationship between government creative industries policy agencies and local cultural producers and entrepreneurs was weakened over time, as narratives of economic development took priority, appearing to key decision makers as more tangible, more achievable and more urgent. In the case of Manchester, this trajectory is traced in the turning away from local cultural agencies such as CIDS towards meta-strategies developed at the national level, such as the proposed relocation of large sections of the British Broadcasting Corporation (BBC) to the newly developed Salford Media City. In an ironic illustration of how policy discourses travel and re-circulate, the arguments for developing Salford as a 'media city' revolved fundamentally around the virtues of developing a media cluster *ex nihilo*, and related opportunities to develop the location as a hub for 'creative class' work force.

Angela Lin Huang considers whether the state-led investment boom in Chinese media, centred on the city of Beijing, can provide the catalyst for Beijing to become an internationally recognised media capital. The concept of 'media capitals' is taken from the work of Michael Curtin (2003, 2007), who uses the term to identify how certain cities develop a self-sustaining momentum to become leading sites of media production and export. Hollywood is obviously the exemplar of this, but cities such as Hong Kong, Mumbai and Cairo have claims to have been regional media capitals, and other cities such as Miami, Seoul and Lagos have been aspiring to such a status. Beijing is the centre of media production in China, with the major media and media training institutions based there, and Huang discusses how the relocation of China Central Television (CCTV) to its new headquarters in the Beijing CBD acted as the basis for the formation of a new media cluster in the Chaoyang District in the city's east. The promotion of CCTV and other key Chinese media institutions is part of the government's plans to promote Chinese 'soft power', or an international cultural influence commensurate with its growing global economic clout, and there has been a major resource commitment to this goal over the 2000s and 2010s. However, Huang identifies several contradictions to this strategy, most notably in the tensions between media continuing to be seen as an ideological apparatus of the communist party-state and the goals of promoting creativity and innovation in media and cultural production. The extent of state control over media products continues to act as a barrier to the development of films and television programs that have audience appeal outside of China and, as a result, the returns on investment that can reverse China's current deficit in cultural trade have been considerably less than was anticipated.

CULTURAL ECONOMY

This subject of creative industries and cities opens up possibilities for a dialogue between disciplines and across bodies of knowledge that have kept some distance from one another. One point of intersection between cultural economic geography and cultural studies. Contributors to this collection such as Chris Gibson and Christy Collis have been engaged in furthering such developments with the institutional support of entities such as the Australian Research Council Cultural Research Network (Gibson 2006), and it is highly likely that such conversations will continue on the context of what Soja has referred to as the 'spatial turn' in social theory. There is also a growing body of work in the field of media geography (Adams 2009; Christophers 2010) and international associations are increasingly developing themes around communication, creative industries and cities, as with the 2011 International Association for Media and Communication Research (IAMCR) conference in Istanbul, Turkey, which had the theme of 'Cities, connectivity, creativity' (http://iamcr2011istanbul.com/?page_id=8).

Another clear potential point of intersection would be between cultural studies and economics. In order to achieve this, however, there are considerable problems with this engagement on both sides, with neo-classical economics possessing an impoverished understanding of cultural dynamics, and cultural studies preferring to critique the caricatured figures of neo-liberalism than engage with more current trends in fields such as institutional, behavioural and evolutionary economics, or with complexity theories (Hartley 2009; Flew 2009). The chapter by Cunningham and Flew considers the degree to which the critical humanities literature on creative industries has in effect walled itself off from a meaningful engagement with economics, by strongly attaching itself to the stalking horse of neo-liberalism as an all-encompassing dominant ideology, rather than displaying more intellectual curiosity about how cultural production and cultural markets may be changing, and meaningful insights than can be derived from the field of economics on such phenomena.

The rise in debates about the creative industries and their wider social and economic significance has been linked to debates about the nature of *cultural economy*. Where the contribution of the arts, media and cultural sectors remained relatively minor in relation to the economy as a whole, it made sense to think about these sectors separately to other form of economic activity. In mainstream economics, this saw cultural economics emerge as a sub-branch of the economics discipline. As Towse (2010: 10–16) observes, much of this work remained at the margins of the field overall, and remained by the lack of 'consensus as to whether the arts are amenable to economic analysis' at all (Towse 2010: 15), although economists as prominent as Adam Smith, W. S. Jevons, John Maynard Keynes, John Kenneth Galbraith and Lionel Robbins all made significant observations about the relationship between culture and economics.

In the critical political economy tradition, cultural work was not seen as 'productive' labour in the sense of directly contributing to the production of value, and cultural sectors were seen as part of the ideological superstructure that acted as a support to the economic 'base' around which the capitalist mode of production was most directly operative. Yet as early as the 1970s, authors such as Raymond Williams were arguing that the distinctions between base and superstructure, and productive and unproductive labor, were increasingly untenable in the face of the massive growth in the global media and communications industries:

> The major modern communications systems are now so evidently key institutions in advanced capitalist societies that they require the same kind of attention … that is given to the institutions of industrial production and distribution … These analyses force theoretical revision of the formula of base and superstructure and of the definition of productive forces, in a social area in which large scale capitalist activity and cultural production are now inseparable. (Williams 1977: 136)

The insistence that culture and economy have come ever closer in the late twentieth and early twenty-first centuries has come from a number of sources. At the empirical level, it is estimated that the combined media, Internet, telecommunications and ICT industries accounted for 6.5 per cent of global GDP in terms of revenue in 2009 (Winseck 2011: 11), and that the creative industries account for between 3 and 10 per cent of national GDP across a diverse range of economies (WIPO 2012). Conceptually, Stuart Hall argued that a characteristic feature of what he and others termed 'New Times' meant that the culture/economy distinction had collapsed as:

> Culture had ceased to be, if it ever was, a decorative addendum to the 'hard world' of production, the icing on the cake of the material world. The word is now as 'material' as the world. Through design, technology and styling, aesthetics has already penetrated the world of modern production. (Hall 1988: 28)

Similar arguments were made by Lash and Urry (1994) when they proposed that in the current phase of global capitalism – what they elsewhere referred to as 'disorganised capitalism' in contrast to the 'organised capitalism' of the early twentieth century heyday of manufacturing – it was now the case that 'the economy is increasingly culturally inflected, and … culture is more and more economically inflected' (Lash and Urry 1994: 64). In more recent literature, it has been observed that 'creative industries ... found favour in the framework of national and urban or regional policy formation, in countries where creativity caught the imagination of politicians and policy-makers who wanted to promote "jobs and GDP" (as the economic development mantra has it)' (Hartley 2005: 5). In such an environment, it is proposed that:

> Cultural studies ... has ceased to maintain itself as different from the commercial and governmental cultures it purports to explain ... Humanists and creatives are talking to business faculties. They're even doing business – trying to grow jobs and GDP in the creative economy for their own cities and regions. (Hartley 2004: 130)

Among those who have sought to rethink the relationship between culture and economy in light of such developments, three approaches can be identified. The first insists upon maintaining a strict conceptual separation between 'culture' and 'economy', even while acknowledging that much cultural activity is commodified and that economic practices have a cultural dimension. The neo-Marxist economic geographer Andrew Sayer (2008) proposes that while the cultural and economic realms increasingly intersect, they need to be considered as analytically separate. To do otherwise, for Sayer, risks collapsing the distinction between intrinsic and instrumental values, or between 'cultural norms and values regarding actions [which] have, at least in part, a deontological character, according to which actions are seen as good or bad in themselves rather than in terms of their moral consequences', and economic activities and processes which 'involve a primarily instrumental orientation; they are ultimately a means to an end, satisfying external goals' (Sayer 2008: 52). O'Connor (2009: 392) developed a similar argument, arguing that 'the market only recognizes consumer preferences, and is indifferent to the specific quality of these; as a result markets 'can only satisfy individual wants as they happen to be'. A distinctive set of cultural values therefore needs to inform the public provision of cultural goods and services 'because they are held to raise the general level of people's abilities to become better people, living a more autonomous and fulfilled life' (O'Connor 2009: 392). Turner (2011: 688–689) argued that to propose that cultural studies can engage in anything other than 'the applied critique of the social and political effects of a market economy' constitutes 'a staggering surrender of the politics of cultural studies to the interests of the market'.

The problems with such accounts are several. At an empirical level, it is clear that large parts of the cultural sphere have economic principles such as return on investment at the core of their *modus operandi*, and a blanket moralism towards such sectors would appear to be misguided. There is the risk of bracketing off the arts and state-supported cultural institutions – such as public broadcasters or government film agencies – on the one hand, and commercial media and entertainment industries on the other, and proclaiming the latter as being worthy of only moral critique. Such an approach is unlikely to take us far in better understanding the cultural economy of cities, where publicly supported and commercial cultures form a complex ecology, as Flew and Cunningham discuss in this volume. Such accounts also upon the theory of neo-liberalism as the dominant ideology of global capitalism, which Flew and Cunningham also critique as being unduly reductionist, as lacking a historical or comparative dimension, and as misinterpreting the work of key thinkers who have used the concept, most notably Michel Foucault (Flew 2012b).

Du Gay and Pryke (2002: 9) describe the sorts of distinctions being made by authors such as Sayer, O'Connor and Turner as being between 'an economic logic that is concerned with the calculation of means/ends relations and a cultural logic that is concerned with ends in themselves', which in turn enables a distinction to be made between the 'instrumentalism' of economic discourse and the intrinsic values and ethical commitments associated with culture, or with cultural studies. They critique this distinction between 'intrinsic and instrumental values', or between formal and substantive rationality to use the terms originally proposed by Max Weber (1978) in considering such questions, 'are too contextually specific and historically contingent to allow them to function as a general "means-ends" moral discriminator' (du Gay and Pryke 2002: 11).

The debates out of which such distinctions arise have a history in Australia, around the role of humanities education in the unified national higher education system established in the late 1980s, and the complementary critique of 'economic rationalism' initiated by the sociologist Michael Pusey (1991) during that period. In both instances, it was argued that instrumentalist reasoning had driven out a prior concern with cultural values and ethical norms. In the case of higher education, this was seen in the rise of a governmental bureaucracy setting policy for universities that was focused upon vocational outcomes and the national economy, while in public policy more generally the rise of a *caste* of economists was seen as having captured the machinery of government policy-making from an earlier generation of policy intellectuals more concerned with nation-building. In both instances, the decision-makers in question are critiqued for their lack of ethical 'wholeness', in a manner very similar to how economics as a discipline is positioned as alterior to culture/cultural studies in the discussions above. From this perspective, the relative lack of influence over decision-makers that cultural critics and humanities intellectuals – and for Pusey, sociologists – may have *vis-à-vis* economists is compensated for by their superior moral and ethical standing.

Hunter (1988, 1992, 1994), Meredyth (1992) and Burchell (1994) all made the point that, from a historical perspective, such accounts were based upon a fantasy of origins. There was simply no prior 'Golden Age' that could be identified as the point from which the contemporary situation would present a moment of moral lapse. Rather, there has been a projection of particular attributes seen as desirable in the fully developed human subject, and, by extension, a 'good society', against which existing institutional arrangements invariably present themselves as inadequate. Hunter argued that cultural studies had in this manner taken on a particular form of 'the discipline of ethical self-shaping' from the traditional humanities – even if as part of a process of claiming to have transcended those traditional humanities disciplines – that has the practical effect of divorcing itself from 'the social machinery in which cultural attributes have been formed as objects of knowledge and administration' (Hunter 1988: 88). For Hunter, this propensity to ignore the complex conditions of existence of cultural institutions leads 'to self-delusion and ends only in moral grandiloquence' (Hunter 1992: 489).

It is notable that such habits of thought continue to be resurrected. In his critique of creative industries policies, Turner (2011) compares them unfavourably to a prior 'Golden Age' of cultural policy studies operating in Australia in the early 1990s. He argues that whereas cultural policy studies was concerned with the public good and 'critique in the public interest', creative industries by contrast has been 'reflecting the classic neo-liberal view that commercial success or "wealth creation" for the enterprises concerned in itself constitutes a public good' (Turner 2011: 692, 693). Yet when we look at the Australian media and cultural policy literature of the early 1990s, what we find are a series of critiques of media policy bureaucrats who have been 'captured' by the logic of economic instrumentalism and fail to fully appreciate the cultural and public interest dimensions of media regulation (Cunningham 1992; Hawke 1995; for an overview, see Flew 2006). So while for Pusey and others critiquing the economic rationalism of the early 1990s, the 1960s and 1970s appeared as a kind of 'Golden Age' where the intellectual inputs were somehow broader, for Turner, writing in the 2010s, the 1980s and early 1990s now appear as a time where public good principles prevailed, before the policy space was colonised by neo-liberal economists and their fellow travellers. Although the timeframes differ, the same propensity to lapsarian accounts remains, based around similar deployments of binary logics.

A second approach to the question of cultural economy draws attention to the increased *culturalisation of the economy*, arguing that the 'cultural turn' has impacted significantly upon how contemporary capitalist economies operate. As noted above, authors such as Hall (1988) and Lash and Urry (1994) have developed this line of argument. George Yúdice (2003) proposed something similar with his concept of *culture as resource*; Yúdice argues that in early twenty-first century global capitalism 'culture is increasingly wielded as a resource for both socio-political and economic amelioration, that is, for increased political participation in an era of waning political involvement, conflicts over citizenship, and the rise of ... "cultural capitalism"' (Yúdice 2003: 9). Such a cultural economy perspective has been further developed in journals such as the *Journal of Cultural Economy*, which describes its principal aim as being to understand 'the role played by various forms of material cultural practice in the organization of the economy and the social, and of the relations between them' (*Journal of Cultural Economy*, Aims and Scope).

In their work on cultural economy, du Gay and Pryke (2002) associated a growing 'culturalisation of the economy' with such factors as:

1. arguments that the management of culture has become the key to improving organisational performance, particularly when there is an alignment between organisational goals and the values and attitudes if those working within them
2. the rise of the service sector, where the relationship between economic transactions and their performative or cultural dimension through interpersonal relations and communicative practices are more overt and visible
3. the rise of the cultural or creative industries, and the spread of practices throughout the economy that have their genesis in these industries, such as a premium being placed upon design, the role of cultural intermediaries in channeling consumer demand, and the role played by networks in time-based and project-based production activities.

Du Gay and Pryke reject the proposition that it is either empirically possible or desirable to distinguish between the economic and cultural realms on the basis of categorical distinctions between 'instrumentally' and 'intrinsically' oriented activities. They follow Max Weber (1978) and Ian Hunter (1988) in arguing that 'formally and substantively oriented conducts are too contextually specific and historically contingent to allow them to function as a general "means-ends" moral discriminator' (du Gay and Pryke 2002: 11). At the same time, the culturalisation of the economy raises the question, as observed by Don Slater, of whether a greater conceptual awareness of the need to think about culture and economy in tandem rather than separately is being conflated with the different empirical claim that cultural factors have become more important in contemporary capitalist economies:

> There is the constant danger of confusing new movements within thought (the new understanding that culture and economy cannot be theorized separately) from new empirical developments. Is it the case that culture is actually more central to economic process than it was before? We need to develop more adequate theories of the sociology of economic life rather than proclaim epochal social revolutions that are merely the artefact of the inadequate theories and theoretical division of labour we have inherited. (Slater 2002: 76)

A third and more radical way of conceiving of the culture/economy relationship arises from *actor-network theory*. Amin and Thrift develop a historical perspective on the concept of cultural economy, arguing that

> The production, distribution and accumulation of resources – loosely the pursuit of prosperity – has always been a cultural performance … [but] with the rise of a separate profession of economics and a set of specifically economic knowledges, such performance has either been neglected or actively denigrated. (Amin and Thrift 2004: xii).

This raises the question of the extent to which the separation of culture and economy is as much a product of discourse as an empirical proposition or one arising from different conceptions of norms and values. Actor-network theory, like post-structuralism in the social sciences, draws attention to the extent to which all of the key categories of economic theory – markets, industries, goods and services – are formed through processes that are simultaneously cultural and economic, as well as being shaped by law, politics, regulations and science. To give one example of how this works in practice, Callon *et al.* (2004) have argued that one of the features of what is sometimes referred to as the 'new economy' is that consumers are themselves increasingly invited to participate in the processes through which one type of product is differentiated from a seemingly similar one on the basis of its perceived 'qualities', but that the qualities do not exist independently of the judgments made by multiple agents, including consumers as well as producers, advertisers, marketers, regulators etc.[1]

There is also the problem that the image of economics and economists constructed here is a phantasm – a projection of certain fears and loathings held by cultural intellectuals onto another academic discipline and body of knowledge. Interestingly, it has been the cultural studies

academic Larry Grossberg who has been most critical of what he describes as intellectual laziness in the engagement of cultural studies with economics. He chides critical humanities scholars only read the work of those economists who they largely agree with, and talk about the other 'discourses of economics as a kind of "ghostly presence" of ideology, and … conflate discourse and practice, intention and effect (as in much of the writing on neoliberalism' (Grossberg 2010: 106). Grossberg argues that:

> The apparent inability or unwillingness to criticize economics as useful knowledge from anything but a radically external position produces an extreme disconnection between socio-cultural criticism and the world of economics. Too often, the criticism of academic economics is founded on an imaginary summation, which is really a relative ignorance, of economics; in addition, the point from which such criticisms are offered is often not a theorized analysis of real economic complexities, but an imagined position of radical opposition, in which the only possible politics is defined by the moral project of overthrowing capitalism. (Grossberg 2010: 107)

Given the growing imbrications of culture and economy that is associated with the rise of the creative industries, among other trends, the urgency of developing a more sustainable, intellectually beneficial and less morally combative account of the relationship between cultural studies and economics is clear. There are positive signs on the horizon. These include work that considers the creative industries in light of institutional and evolutionary economics (Potts 2011; Hartley *et al.* 2013 (forthcoming)), the work on cultural economy cited above as well as work that critically engages with economic discourses (Ruccio 2008; Grossberg 2010), and engagement with new developments in the natural sciences such as complexity theories and their implications for cultural research (Ang 2011; Sofoulis 2011). This collection of essays is also aimed at making a modest contribution to such ongoing conversations.

NOTES

1 An example of how an object can be defined through policy discourses can be found with the Internet. In the United States, the striking down of the *Communications Decency Act* in the U.S. Supreme Court was premised upon the idea of the Internet being 'like print' and therefore subject to the provisions of the U.S. Constitution guaranteeing freedom of speech. By contrast, in Australia the provisions of the *Broadcasting Services Act 1992* were extended to the Internet by the Howard government in 1997 through an *Online Services Amendment*, which in effect defines the Internet as being 'like television' for regulatory purposes.

REFERENCES

Adams, P. 2009. *Geographies of media and communication*. Malden, MA: Wiley-Blackwell.

Amin, A. 2003. Industrial districts. In *A companion to economic geography*, eds E. Sheppard and T. Barnes, 149–68. Oxford: Blackwell.

Amin, A. and N. Thrift. 2004. Introduction. In *The Blackwell cultural economy reader*, eds A. Amin and N. Thrift, x-xxx. Oxford: Blackwell.

Ang, I. 2011. Navigating complexity: From cultural critique to cultural intelligence. *Continuum: Journal of Media and Cultural Studies* 25(6): 779–94.

Bassett, K., I. Smith, M. Banks and J. O'Connor. 2005. Urban dilemmas of competition and cohesion in cultural policy. In *Changing cities: Rethinking urban competitiveness, cohesion and governance*, eds N. Buck, I. Gordon, A. Harding and I. Turok, 132–53. Basingstoke: Palgrave Macmillan.

Burchell, D. 1994. *Economic government and social science: The economic rationalism debate*. Cultural Policy Paper No. 1. Brisbane: Institute of Cultural Policy Studies, Griffith University.

Cairncross, F. 1998. *The death of distance: how the communications revolution will change our lives*. London: Orion Business Books.

Callon, M., C. Méadel and V. Rabeharisoa. 2004. The economy of qualities. In *The Blackwell cultural economy reader*, eds A. Amin and N. Thrift, 58–79. Oxford: Blackwell.

Castells, M. 1996. *The rise of the network society*. Oxford: Blackwell.

Caves, R. 2000. *Creative industries: Contracts between art and commerce*. Cambridge, MA: Harvard University Press.

Chapain, C. and R. Communian. 2010. Enabling or inhibiting the creative economy? The role of the local and regional dimensions in England. *Regional Studies* 44(6): 717–34.

Clapson, M. 2003. *Suburban century: Social change and urban growth in England and the USA*. Oxford: Berg.

Christophers, B. 2009. *Envisioning media power: On capital and geographies of television*. Lanham, MD: Lexington Books.

Collis, C., S. Freebody and T. Flew. 2012. Seeing the outer suburbs: Addressing the urban bias in creative place thinking. *Regional Studies* (forthcoming).

Cooke, P. and L. Lazzeretti. 2008. Creative cities: An introduction. In *Creative cities, cultural clusters and local economic development*, eds P. Cooke and L. Lazzeretti, 1–22. Cheltenham: Edward Elgar.

Cunningham, S. 1992. *Framing culture: Criticism and policy in Australia*. Sydney: Allen & Unwin.

Currid, E. 2007. *The Warhol economy: How fashion, art and music drive New York City*. Princeton, NJ: Princeton University Press.

Curtin, M. 2003. Media capital: Towards the study of spatial flows. *International Journal of Cultural Studies*, 6(2): 202–28.

Curtin, M. 2007. *Playing to the world's biggest audience: The globalization of Chinese film and TV*. Berkeley, CA: University of California Press.

Drake, G. 2003. 'This place gives me space': Place and creativity in the creative industries. *Geoforum* 34(4): 511–24.

du Gay, P. and M. Pryke. 2003. Cultural economy: An introduction. In *Cultural economy: Cultural analysis and commercial life*, eds P. du Gay and M. Pryke, 1–19. London: Sage.

Evans, G. 2009. Creative cities, creative spaces and urban policy. *Urban Studies* 46(6): 1003–40.

Flew, T. 2006. The social contract and beyond in broadcast media policy. *Television and New Media* 7(3): 282–305.

Flew, T. 2009. The cultural economy moment? *Cultural Science* 2(1). http://cultural-science.org/journal/index.php/culturalscience/article/view/23/79 (accessed 15 November, 2009).

Flew, T. 2011. Right to the city, desire for the suburb? *M/C Journal* 14(4). http://journal.media-culture.org.au/index.php/mcjournal/article/viewArticle/368 ().

Flew, T. 2012a. *The creative industries, culture and policy*. London: Sage.

Flew, T. 2012b. 'Michel Foucault's *The Birth of Biopolitics* and contemporary neo-liberalism debates', *Thesis Eleven: Critical Theory and Historical Sociology*: 108(1): 42–63.

Florida, R. 2002. *The rise of the creative class*. Cambridge, MA: Basic Books.

Florida, R. 2008. *Who's your city?* New York: Basic Books.

Galloway, S. and S. Dunlop. 2008. A critique of definitions of cultural and creative industries in public policy. *International Journal of Cultural Policy* 13(1): 17–31.

Garnham, N. 2005. An analysis of the implications of the 'creative industries' approach to arts and media policy making in the United Kingdom. *International Journal of Cultural Policy* 11(1): 15–29.

Gertler, M. 2003. A cultural economic geography of production. In *Handbook of cultural geography*, eds K. Anderson, M. Domosh, S. Pile and N. Thrift, 131–46. London: Sage.

Gibson, C. 2006. The Cultural Research Network: Opportunities for a rhizomic future for geography in Australia? *Geographical Research* 44(4): 418–21.

Gibson, C. 2010. Creative geographies: Tales from the 'margins'. *Australian Geographer* 41(1): 1–10.

Gibson-Graham, J. 2000. Poststructural interventions. In *A companion to economic geography*, eds E. Sheppard and T. J. Barnes, 95–110. Malden, MA: Blackwell.

Glaeser, E. 2011. *Triumph of the city*. London: Macmillan.

Gordon, I. and P. McCann. 2001. Industrial clusters: Complexes, agglomeration, and/or social networks? *Urban Studies* 37(3): 513–32.

Grossberg, L. 2010. *Cultural studies in the future tense*. Durham, NC: Duke University Press.

Hall, P. 1998. *Cities in civilization: Culture, innovation and urban order*. London: Phoenix Grant.

Hall, S. 1988. Brave new world. *Marxism Today*. October: 24–28.

Hartley, J. 2004. The 'value chain of meaning' and the new economy. *International Journal of Cultural Studies* 7(1): 129–40.

Hartley, J. 2005. Creative industries. In *Creative industries*, ed. J. Hartley, 1–40. Oxford: Blackwell.

Hartley, J. 2009. From cultural studies to cultural science. *Cultural Science* 2(1). http://cultural-science.org/journal/index.php/culturalscience/article/view/19/68 (accessed 16 November, 2009).

Hartley, J., J. Potts, S. Cunningham, T. Flew, M. Keane and J. Banks. 2013 (forthcoming). *Key concepts in the creative industries*. London: Sage.

Harvey, D. 1982. *The limits to capital*. Oxford: Blackwell.

Harvey, D. 2008. The right to the city. *New Left Review* 53 (Sept–Oct): 23–40.

Hawke, J. 1995. Privatising the public interest: The public and the *Broadcasting Services Act* 1992. In *Public voices, private interests: Australia's media policy*, eds J. Craik, J. J. Bailey and A. Moran, 33–50. Sydney: Allen & Unwin.

Hunter, I. 1988. *Culture and government: The emergence of literary education*. London: Macmillan.

Hunter, I. 1992. The humanities without humanism. *Meanjin* 51(3): 479–90.

Hunter, I. 1994. *Rethinking the school: Subjectivity, bureaucracy, criticism*. Sydney: Allen & Unwin.

James, A., R. Martin and P. Sunley. 2008. The rise of cultural economic geography. In *Economic geography: Critical concepts in the social sciences – Volume IV: The cultural economy*, eds R. L. Martin and P. Sunley, 3–18. London: Routledge.

Johnson, L. 2012. Creative suburbs: How women, design and technology renew Australian suburbs. *International Journal of Cultural Studies*, 217–29.

Keane, M. 2009. The capital complex: Beijing's new creative clusters. In *Creative economies, creative cities: Asian-European perspectives*, eds L. Kong and J. O'Connor, 77–95. Dordrecht: Springer.

Kong, L. 2009. Beyond networks and relations: Towards rethinking creative cluster theory. In *Creative economies, creative cities: Asian-European perspectives*, eds L. Kong and J. O'Connor, 61–76. Dordrecht: Springer.

Kong, L., C. Gibson, L. M. Khoo and A. L. Semple. 2006. Knowledges of the creative economy: Towards a relational geography of diffusion and adaptation in Asia. *Asia Pacific Viewpoint* 47(2): 173–94.

Kotkin, J. 2005. *The new suburbanism: A realist's guide to the American future*. http://www.joelkotkin.com/Urban_Affairs/The%20New%20Suburbanism.pdf (accessed March 20, 2009).

Kotkin, J. 2007. What is the new suburbanism? In *Planetizen: Contemporary debates in urban planning*, eds A. Chavan, C. Peralta and C. Steens, 28–33. Washington, DC: Island Press.

Landry, C. 2000. *The creative city: A toolkit for urban innovators*. London: Earthscan.

Landry, C. 2005. London as a creative city. In *Creative industries*, ed. J. Hartley, 233–43. Oxford: Blackwell.

Lash, S. and J. Urry. 1994. *Economies of signs and space*. London: Sage.

Latour, B. 2007. *Reassembling the social: An introduction to actor-network theory*. Oxford: Oxford University Press.

Lorenzen, M., and L. Frederiksen. 2008. Why do cultural industries cluster? Localization, urbanization, products and projects. In *Creative cities, cultural clusters and local economic development*, eds P. Cooke and L. Lazzeretti, 155–79. Cheltenham, UK: Edward Elgar.

Marshall, A. 1990 [first published 1890]. *Principles of economics*. 8th edn. London: Macmillan.

Martin, R. and P. Sunley 2003. Deconstructing clusters: Chaotic concept or policy panacea? *Journal of Economic Geography*, 3(1): 3–35.

Meredyth, D. 1992. Changing minds: Cultural criticism and the problem of principle. *Meanjin* 51(3): 491–504.

Mommaas, H. 2009. Spaces of culture and economy: Mapping the cultural-creative cluster landscape. In *Creative economies, creative cities: Asian-European perspectives*, eds L. Kong and J. O'Connor, 45–60. Dordrecht: Springer.

Oakley, K. 2004. Not so cool Britannia: The role of the creative industries in economic development. *International Journal of Cultural Studies* 7(1): 67–77.

O'Connor, J. 2009. Creative industries: A new direction? *International Journal of Cultural Policy* 15(4): 387–402.

Palmer, R and G. Richards. 2011. *Third European cultural capital report 2011*. Arnhem: ATLAS.

Peck, J. 2005. Struggling with the creative class. *International Journal of Urban and Regional Research* 29(4): 740–70.

Perez, C. 2010. Technological revolutions and techno-economic paradigms. *Cambridge Journal of Economics* 34(1): 185–202.

Porter, M. E. 1998. Clusters and the new economics of competitiveness. *Harvard Business Review* 76: 77–90.

Porter, M. E. 2000. Location, competition, and economic development: Local clusters in a global economy. *Economic Development Quarterly* 14(1): 15–34.

Potts, J. 2011. *Creative industries and economic evolution*. Cheltenham, UK: Edward Elgar.

Pratt, A. 2008. Creative cities: The cultural industries and the creative class. *Geografiska Annaler: Series B – Human Geography* 90(2): 107–17.

Pratt, A. 2009. Policy transfer and the field of cultural and creative industries; What can be learned from Europe? In *Creative economies, creative cities: Asian-European perspectives*, eds L. Kong and J. O'Connor, 9–24. Dordrecht, Holland: Springer Press.

Pusey, M. 1991. *Economic rationalism in Canberra: A nation-building state changes its mind*. Melbourne: Cambridge University Press.

Ross, A. 2007. Nice work if you can get it: The mercurial career of creative industries policy. *Work Organisation, Labour, & Globalisation* 1(1): 1–19.

Ruccio, D. 2008. Economic representations: What's at stake? *Cultural Studies* 25(6): 892–912.

Sayer, A. 2008. The dialectic of culture and economy. In *Economic geography: Critical concepts in the social sciences – Volume 4: The cultural economy*, eds R. L. Martin and P. J. Sunley, 50–68. London: Routledge.

Schuster, J. M. 2002. Sub-national cultural policy – where the action is? Mapping State cultural policy in the United States. *International Journal of Cultural Policy* 8(2): 181–96.

Scott, A. J., J. Agnew, E. Soja, and M. Storper. 2001. Global city-regions. In *Global city-regions: trends, theory, policy*, ed. A. J. Scott, 11–30. Oxford: Oxford University Press.

Scott, A. J. 2008. *Social economy of the metropolis: cognitive-cultural capitalism and the global resurgence of cities*. Oxford: Oxford University Press.

Slater, D. 2002. Capturing markets from the economists. In *Cultural economy: Cultural analysis and commercial life*, eds P. du Gay and M. Pryke, 147–63. London: Sage.

Sofoulis, Z. 2011. Skirting complexity: The retarding quest for the average water user. *Continuum: Journal of Media and Cultural Studies* 25(6): 795–810.

Soja, E. 2010. *Seeking spatial justice*. Minneapolis: University of Minnesota Press.

Stevenson, D. 2004. 'Civic gold' rush: Cultural planning and the politics of the 'third way'. *International Journal of Cultural Policy* 10(1): 119–31.

Storper, M. and A. Scott. 2009. Rethinking human capital, creativity and urban growth. *Journal of Economic Geography* 9(1): 147–67.

Thrift, N. 1999. Capitalism's cultural turn. In *Culture and economy after the cultural turn*, eds L. Ray and A. Sayer, 135–61. London: Sage.

Thrift, N. 2005. *Knowing capitalism*. London: Sage.

Towse, R. 2010. *A textbook of cultural economics*. Cambridge: Cambridge University Press.

Turner, G. 2011. Surrendering the space: Convergence, culture, cultural studies, and the curriculum. *Cultural Studies* 25(4/5): 685–99.

Weber, M. 1978. *Economy and society: An outline of interpretative sociology* (2 vols). Berkeley, CA: University of California Press.

Williams, R. 1977. *Marxism and literature*. Oxford: Oxford University Press.

Winseck, D. 2011. The political economy of the media and the transformation of the global media industries. In *Political economies of the media: The transformation of the global media industries,* eds D. Winseck and D. Y. Jin, 3–44. London: Bloomsbury.

World Intellectual Property Organisation (WIPO). 2012. *Creative industries – economic contribution and mapping.* http://www.wipo.int/ip-development/en/creative_industry/economic_contribution.html (accessed 16 February, 2012).

Worldwatch Institute. 2007. *State of the world 2007: Our urban future.* Washington, DC: Worldwatch Institute.

Yúdice, G. 2003. *The experience of culture: Uses of culture in the global age*. Durham, NC: Duke University Press.

Beyond the Inner City: Real and Imagined Places in Creative Place Policy and Practice

Christy Collis and Emma Felton

Creative Industries Faculty, Queensland University of Technology, Brisbane, Queensland, Australia

Phil Graham

Institute for Creative Industries and Innovation, Queensland University of Technology, Brisbane, Queensland, Australia

As the economic and social benefits of creative industries development become increasingly visible, policymakers worldwide are working to create policy drivers to ensure that certain places become or remain "creative places." Richard Florida's work has become particularly influential among policymakers, as has Charles Landry's. But as the first wave of creative industrial policy development and implementation wanes, important questions are emerging. It is by now clear that an "ideal creative place" has arisen from creative industries policy and planning literature and that this ideal place is located in inner cities. In this article, the authors shift focus away from the inner city to where most Australians live: the outer suburbs. They report on a qualitative research study into the practices of outer-suburban creative industries workers in Redcliffe, Australia, arguing that the accepted geography of creative places requires some recalibration once the material and experiential aspects of creative places are taken into account.

This article presents preliminary findings from an Australian research project called *Creative Suburbia*. It argues that "creative place" policy, planning, and analysis literatures tend to be based on a separation between the imagined aspects of place from its material and experiential aspects. Policy imaginations informed by "creative city" strategies (Florida 2002) may lead to a mistaken, "one-size-fits-all" emphasis on inner-city locations as the focus for creative industries workers, creative clusters, and "creative place" policy. Conversely, the suburbs are construed as unproductive, passive, and culturally moribund. As our preliminary findings demonstrate, not only are creative industries active in outer suburbs, but interview responses also consistently foreground the gaps between policy imaginings of inner-city creative places and the lived experience of outer-suburban materialities for creative workers.

As the economic and social benefits of creative industrial development become increasingly visible, local planning authorities are responsive to "creative place" ideas and policy drivers. Meric S. Gertler (2004, 1) notes that "with the widely acknowledged shift to a knowledge-based . . . economy, creative cities have become the key locus for the creation of economic value." This article draws on the work of Chris Gibson and Chris Brennan-Horley (2006), which critiques the emphasis on inner-urban localities in creative place thinking in Australia and argues for a more complex understanding of where and how the creative industries operate beyond inner-urban areas.

First, this article reviews a selection of current creative place discourses to show how they privilege inner-urban sites. Second, it presents early findings of qualitative research with outer suburban creative industries workers in an Australian outer suburban locality. This article uses the Queensland State Government's definition of creative industries, which includes "music composition and production; film, television, and entertainment software; writing, publishing, and print media; advertising, graphic design, and marketing; architecture, visual arts, and design" (Queensland 2008; see also Higgs et al. 2007). For research grounded in spatial experience, we use Edward Soja's (1996) understanding of "place" as being comprised of three mutually dependent aspects: an objective material space, the ways in which space is imagined and represented, and the ways in which it is experienced by people. Place is most coherent when all three of these constituent elements are in alignment (Lefebvre 1992). This type of geographical approach to studies of the creative industries remains uncommon (Gibson and Connell 2003; Gibson, Murphy, and Freestone 2002). Creative place research that focuses on all three aspects of place provides a complex and nuanced understanding of the ways in which creative industries operate in their outer suburban localities.

IMAGINED PLACE: PLACE IN CREATIVE INDUSTRY DISCOURSE

Creative city policies and analyses are proliferating across a range of geographical sites: from international developments such as UNESCO's Creative Cities Network to national strategies such as Singapore's 2002 Creative Industries Development Strategy. At the local scale, there are city-wide strategies such as Landry's 2003 "Rethinking Adelaide" report, and highly local "creative place" measurement indices such as those used in the Australian Local Government Association's "State of the Regions" analysis (National Economics 2002). Florida's work (2002, 2005) has achieved canonical status among policymakers, as has Landry's (2000). But as what might be termed the first wave of creative city policy development and implementation wanes, important questions are surfacing.

In particular, questions are emerging about the geographical sites of creative industries, partly produced by a reassertion of place and its social, cultural, and economic role in creative industries development. This is in contrast to an earlier view in which technology was seen by some critics (e.g., Cairncross 1998; Castells 1989) as the enabling vector that would eventually help transcend geographical boundaries and cultural contexts to create a nongeographical "space of flows" for technology-mediated communication (Castells 1989). But as Hutton (2006, 1839) notes, "Place is demonstrably a defining feature of the new production economy of the 21st century ... and 'place' in this context ineluctably comprises both concrete and representational features." Predictions about the death of geography have proven to be incorrect in three ways.

First, drawing on Porter's economic geographical studies of industry clusters, creative industries analysts and planners have become increasingly aware of the existence and the value of "creative clusters": sites in which sectors of the creative industries benefit economically and professionally by close geographical collocation (Porter 2000; cf. Pratt et al. 2007). The Australian Government "Creative Industries Cluster Study" (DCITA 2002), for example, focused specific attention on geography, noting that "cluster approaches potentially offer a means of addressing barriers and market failures for sectors producing digital content and applications" (ibid., 1).

Second, Florida (2005, 37) notes that high-tech workers seem to prefer to live and work in places with specific local characteristics, particularly "technology, talent and tolerance." Florida investigated sites in which high-tech creative industries were flourishing and found that these places shared certain physical and cultural geographical commonalities, such as inclusivity and open social and ethnic diversity, "bohemian enclaves" (ibid., 117), and "natural features and amenities" (ibid., 172). Florida concludes that for creative industries development, geography does matter. Based on his findings, he creates indices with which places could measure their own potential to attract "the creative class" and thus to prosper.

Third, creative industries analysts argue that the cultural characteristics of specific places do not just attract creative workers but trigger particular kinds of creativity (Drake 2003, 513; Helbrecht 2001; Hutton 2006) or endow those places' products with a form of geographical cultural capital (Scott 1999). Geography is thus far from dead when it comes to the creative industries. In fact, place has emerged in the literature as one of the key drivers of creative industrial strength. But what kinds of geography have emerged in creative place thinking as ideal, and as less than ideal, sites for creative industries productivity and development?

The key geographical division which runs through the bulk of creative place analysis and planning is the splitting of cities into "creative" inner cities and "uncreative" suburbs, particularly outer suburbs (Gibson and Brennan-Horley 2006). In this thinking, clustered creative industrial productivity takes place in inner cities, while outer suburbs are "hinterland" sites of uncreative, conservative, dispersed nonproductivity and consumption (Florida 2004). Gibson and Brennan-Horley (2006, 456) characterize this binary as "densely populated vs. sprawl; gentrified terraces and apartment culture vs. new estates and first home buyers; zones of (male) production and creativity against (female) sedate, consumer territory." Florida (2005, 65) characterizes the outer suburbs—which he terms "sprawl"—as a major negative factor in his creative place index. In his analysis, cities such as Los Angeles, San Diego, and Phoenix are awarded "dishonourable mention" due to their high scores on the "sprawl risk ranking" index. When it comes to being a site for creative industrial productivity, Florida (ibid., 64) concludes that "sprawl is a vexing problem." For Florida, there is a direct equation: The more suburbs a city has, the less creative potential it has. In Florida's (2004) analysis, outer suburbs are not just uncreative: in reducing a place's "creative place status," they are actively anticreative, and economically "choking."

When the Australian Local Government Association used Florida's "creative place" indices to determine which places in Australia were the most creative, and therefore most likely "to be successful in the modern globalised economy" (National Economics 2002, i), the winners were (in order) Sydney, Inner Melbourne, the Australian Capital Territory, Central Perth, Central Adelaide, and Brisbane City. Not surprisingly, Australia's outer suburbs, or "sprawl," ranked poorly. Similarly, the National Institute of Economic and Industry Research adopted Florida's indices to rank regions across Australia and compared them to U.S. regions (National Institute 2004). Central Melbourne and Sydney both scored a ranking of 4 (only three U.S. cities ranked higher). But when their outer suburbs were factored in, Sydney's ranking dropped to 26, and Melbourne's to 34 (National Institute 2004). The global uptake of Florida's indices for "creative place" analysis ensures that outer suburbs will never achieve official "creative place" status and are thus unlikely to attract government or business creative industrial investment. As Danaher (2007) notes, in this "metrocentric" vision of creative places, suburbs "can only be poor relations, inadequately seeking to replicate the conditions that enable creative practice to flourish in the cities" (13–14).

Critiques of creative city policy observe that Florida's and Landry's analyses have swiftly been translated into homogenized, prescriptive geographies. Steven Malanga (2004), for example, states that "a generation of leftish policy-makers and urban planners is rushing to implement Florida's vision [just as] an admiring host of uncritical journalists touts it." Kate Oakley (2004, 71) argues that Florida's work has given rise to a one-size-fits-all approach to economic development, with creatively "backward" regions trying desperately—often in the face of all geographical, economic, and social realities—to make themselves look like London, Boston, or San Francisco.

Andrew Ross similarly argues that "creative city" policy tends to take a "cookie cutter approach to economic development [that] does violence to regional specificity" (2006–7, 2). In Australia, Gibson and Lily Kong (2005, 552) observe that

> the emerging model of academic knowledge-to-policy traffic is one of assuming singular "recipes" for success in transforming places based on advice from experts and advisors not well enough grounded in places to account for the more complex and contested geographies they contain.

In Soja's framework, this general effect can be described as a tendency for policy to separate material and experiential geographies from imagined ones, thus giving only a partial and very limited view of creative place. Materiality and experience are devalued in a Taylorist discourse that assumes there is "one best way" (Kanigel 1997) to develop creative places, despite the diversity of materiality, experience, and culture that characterizes any city, town, or suburb that might find itself the object of "creative cities" policy.

To close the gap between the imagined places of creative city policy and the material and experiential geographies of specific sites, it is necessary to understand the significant economic and social shifts that have occurred across Australian metropolitan regions in the past two decades, particularly in its middle and outer suburban areas. This includes an analysis of the ways in which suburban sites have been profoundly transformed, realigning and contesting old assumptions about the metropolitan inner-outer divide.

MATERIAL PLACE: THE CHANGING SUBURBAN LANDSCAPE

The "imagined geography" of outer suburbs generally sees such places as dull sites of domestic consumption rather than creative productivity. However, scholars in the United States (Kotkin 2005) and Australia (Gibson and Brennan-Horley 2006; Gleeson 2002; Randolph 2004) provide evidence of deep structural change in suburban localities. Joel Kotkin's "new suburbanism" thesis in the United States suggests that the suburbs play a crucial role in the dynamism and development of American cities and American culture. Kotkin argues that the "rebirth" of the inner core of American cities has been generally overstated:

> Since 1950 more than 90 per cent of all the growth in U.S. metropolitan areas has been in the suburbs. Most of the fastest growing "cities" of the late twentieth century—Los Angeles, Atlanta, Orlando, Phoenix, Houston, Dallas and Charlotte—are primarily collections of suburbs, often with only marginal links to the traditional urban core. (2005, 9)

Gleeson (2002) similarly argues that "the suburbs of our major cities are the crucible of Australian life, but they are poorly understood and their dynamism is often not appreciated" (2002, 229).

The division between inner city and the suburbs draws upon a long-established discourse in Australia about the suburbs being culturally barren places where nothing of much interest happens. Summing up the deep current of antisuburbanism among Australian intellectuals and artists since white settlement, Garry Kinnane (1998, 42) observes that the suburbs have come to stand for a "living death of conformity and safety," with "'suburbia' denoting a generalised place in the imagination, rather than an actual geographical place." Once described as "the first suburban nation" (Horne 1964), Australian suburban life has long been imagined and experienced as distinct from inner-city life. The association of inner-urban life with excitement, diversity, and inclusivity contrasts sharply with suburban life with its associations of peace, order, and privacy (on the one hand), and homogeneity and desolation (on the other). In Australia, the "imaginary topography" of the outer suburbs is almost entirely negative (Kinnane 1998). Despite material shifts in the geography of the suburbs, an intellectual ambivalence toward the suburbs remains.

Since urban consolidation began in the late 1970s, the Australian suburbs have changed dramatically. One of the most significant shifts in the suburbs is demographic. No longer the heartland of the traditional nuclear family with mum and dad living on a quarter-acre block, in the twenty-first century the suburban population is more atomized, older, and the structure of household composition has changed significantly (Randolph 2004, 484). In the suburbs, couple-only households, single-parent households, and single-person households have all increased rapidly in the ten-year census period from 1996 to 2006 to the extent that they now outnumber households of couples with children (Australian Bureau of Statistics 2006). Australia's multicultural policies and the processes of gentrification have contributed to another major demographic shift in which many outer suburbs have become more ethnically diverse than some inner-city localities. In Brisbane, one of Australia's fastest growing cities, demographic trends in the inner-urban regions reveal a decreasing proportion of ethnic and socioeconomic diversity (Stimson 1998, 211; Australian Bureau of Statistics 2001). Consequently, sites of disadvantage once associated with the inner city and public housing are now emerging in private housing in the middle suburbs (Randolph 2004, 488). The increasing homogeneity of Australia's inner-urban areas is in danger of producing fewer of Florida's creative city requisite markers of diversity, from which (it is assumed) tolerance and inclusivity follow.

The ways in which people live in the outer suburbs has also changed: The quarter-acre block is now being carved up to make way for townhouses, villas, and apartments, producing suburban neighborhoods of high density (Randolph 2004, 489). Higher density housing contributes to an environment with characteristics formerly typical of inner-city areas. Kevin O'Connor and Robert J. Stimson (2004, 45) note a further significant reorganization of material space in suburbia: the proliferation of commercial and public facilities across the breadth of the metropolitan areas, particularly in middle and outer suburbs:

> This outcome is the result of the construction of new hospitals, universities and other community facilities as well as factories, shopping centres, offices and warehouses, restaurants and entertainment facilities which have been built by firms recognising the opportunities created by the re-organisation and re-direction of business operations.

Greater commercial and public infrastructure reflects changes in work-space relations over the past two decades and the complex role of the middle to outer suburbs in the wider economy. The suburbs are no longer passive places from which predominantly male workers leave to commute to the inner city each day. They now support a wider range of occupations and activities, with only 20–30 percent of jobs now located in inner-urban areas (Gipps et al. 1996).

Empirical studies of creative industries activity in Sydney and its surrounding exurban areas present confronting evidence against an inner-city bias: exurban areas such as Wollongong and the Blue Mountains experienced the highest rates of creative industries employment growth in all of Sydney between 1991 and 2001 (Gibson and Brennan-Horley 2006, 467). Similarly, the exurban areas of Wyong, Camden, and Wollondilly achieved higher rates of growth in creative work in the last twenty years than did inner-city areas such as Sydney City and Marrickville (ibid., 465). Yet this finding is only surprising when pitted against the entrenched imagined geography of suburbia. Gibson and Brennan-Horley (ibid., 468) explain that the growth pattern of creative industries in the outer suburbs of Sydney is entirely logical. Outer suburban areas are the fastest growing areas in Australia, so it makes sense that their creative industries workforces are also growing. The growth of creative industries in the outer suburbs also makes sense when paired with real estate trends: For many creative workers, the inner city has simply become too expensive a place in which to work and live so they have moved out to the more affordable outer suburbs.

Empirical evidence increasingly demonstrates that there is a notable level of creative industries activity taking place in outer suburbia and periurban regions. The transformed suburbs are now places of intense commercial and demographic complexity, with some suburbs becoming urban centers in their own right. To understand the ways in which the many material and demographic shifts impact on creative industries in the outer suburbs, it is necessary to address the experiential aspects of outer-suburban place as they are experienced by creative industries workers. For this reason, the Australian Research Council-funded project Creative Suburbia explores creative industries workers principally from an experiential perspective.

EXPERIENTIAL PLACE: EVIDENCE FROM CREATIVE WORKERS IN AN OUTER BRISBANE SUBURB

The Creative Suburbia project is designed to investigate whether, how, and to what extent creative industries are active in outer-suburban areas of two major Australian cities: Brisbane and Melbourne. The research methodology is primarily qualitative, using interviews with creative industries workers based in selected outer suburbs. The results presented here are drawn from interviews with participants who live in Redcliffe, an exurban satellite of Brisbane. Brisbane has experienced significant local and state government investment in cultural infrastructure from the late 1980s, a trend accelerated by urban renewal programs throughout the 1990s (Stimson 1998). This has resulted in several large-scale cultural venues and exhibition spaces around which many smaller scale, privately owned venues and exhibition spaces have been built. All these facilities are within a two-kilometer radius of Brisbane's inner-city core.

Respondents represent a cross section of creative industries workers, including multimedia developers, architects, graphic designers, musicians, and visual artists, all of whom live on the Redcliffe peninsula about 40 km from Brisbane's center. The majority of people work where they live, while a few commute part-time to inner-city workplaces. The most pronounced theme at this stage of the investigation is the respondents' relationship with their suburban location to their work. We group the findings into three key areas in relation to the creative industries professionals and their locational choice of the outer suburbs: (1) economic factors such as affordability; (2) locality preferences for creative workers, which include the concept of creative stimuli (Drake 2003); and (3) professional and creative networks.

Locality and Affordability

Urban renewal occurring in Brisbane's inner city for the last decade has had the well-documented effect of displacing many low-income residents (Zukin 1995). Affordability of location was a constant theme among interview respondents. While many respondents had lived in Redcliffe for many years, a significant number had also recently moved from inner-city locations. Several attributes of locality are identified by respondents as underpinning decisions to live and work in the locality; these varied according to the individuals and to the type of creative industries work in which they were engaged. Primarily, location was perceived as important in terms of affordability of business premises, or of housing, in comparison to inner-city areas (see also Gibson and Brennan-Horley 2006; Gibson, Murphy, and Freestone 2002, 186).

John, a graphic designer and multimedia developer who runs his own business and employs eight staff members, says:

> It's a good place to work because the rent is affordable, it's about half to a third of what you'd pay in the city... it's nice and relaxing here too. I've thought about moving but probably wouldn't because it's easier here.

Affordability appears to work for creative workers and their clients. John also states that low rent is an advantage for his clients because he can keep costs down and his clients know they are not paying extra for his services to cover high rental overheads. This was repeated by Gary, an advertising executive who had recently relocated his office from the inner city. In the event of other inner-city-based advertising executives being forced to put up their fees due to rising rents, he was able to maintain a competitive edge and keep his costs stable. Distance from the city is generally not seen as a problem among respondents with city-based clients. For instance, while many of John's clients are within driving distance of the premises and the majority (about 65 percent) are local and regionally based, he does not always meet them and many remain "virtual" clients. Similarly, while Gary had maintained his predominantly city-based clients, he either deals with them online or is happy to drive into the city for occasional meetings. The flexibility provided by online technologies and the relative ease of commuting by car (but not public transport) enable a flow between the city and outer suburbs that, in the case of our respondents, meets both their clients' and their own needs.

While technology has not rendered geography redundant as widely predicted, it has contributed to the flexibility and reach of certain work practices and markets (Drake 2003, 512). This was evident in the case of Declan, a graphic designer who works part-time in the inner city and part-time at home; he runs an online T-shirt design business selling to clients locally, nationally, and worldwide from the latter.

Most respondents in the twenty-five-to-thirty-nine years age group moved to Redcliffe from the inner city because they couldn't afford to purchase inner-city houses. Although several respondents stated that they once would have preferred to buy homes in the inner city and had misgivings about moving to an outer suburb, they had little choice if they were to purchase their own home. However, once they had made the move, all were unequivocal about their decision. Lifestyle factors that combine work and family in an environment that is both financially and culturally sustaining were cited as significant reasons to live and work in an outer suburb. A third of visual artists and musicians identified affordability as one of the main reasons for moving to Redcliffe, with most of these having studios in or under their houses.

Locality and Creative Work

Environmental influences in relation to creativity have been explored in psychological and other literature (Drake 2003; Menin and Samuel 2003; Lubart and Sternberg 1998). For our purposes here we focus on the specificity of the outer-suburban experience of creativity compared to

19

that of the inner city. Responses to the question of whether and to what extent their suburban environment influenced respondents' creative work are spread along a continuum from "not at all" to far greater elaborations of the ways in which locality is critical to their creative work and business.

For some respondents, lifestyle is also integral to productivity and the outer suburbs are seen as facilitating a balance between work and play. Although negative factors were identified—such as a smaller local market for their products and services and clients who were not willing to pay as much as a city-based business could expect—the positive attributes of location were couched more in terms of a psychosocial relationship to place that has an impact on the creative process itself (see also Drake 2003). Experiential articulations of place and creativity were frequently cited, with serenity and freedom from distraction being one of the key positive attributes of the locality. This is a theme dominant among musicians and visual artists. Forty-three percent of interview respondents described their attachment in physical terms, particularly in relation to the geographical separation of the Redcliffe peninsula from the mainland. Larry, a visual artist, describes the peninsula as "an island surrounded by a moat," a "haven," and a "sanctuary." Leah, a sculptor, identified the area as free from distractions and compares it to where she formerly lived in Brisbane's inner city: "It's far less distracting . . . in Brisbane there were too many distractions, too many exhibition openings and things to go to." Contrary to Florida's conceptions, creative industries workers may not necessarily gravitate towards buzzing inner-urban hubs.

Several respondents who had moved to the suburb from areas they considered more creative expressed surprise at the ways in which the locality was able to support their creative endeavors in a variety of ways. Julie, a fashion designer and retailer, moved from a northern coastal town with a strong arts-based community. She was pleasantly surprised at how the locality was "ready for something different," and that

> it's encouraged me to be more creative because they are willing to accept things that are different . . . it's given me the incentive to actually get back into being more creative.

Julie later muses that although in some ways she would have liked her business to be based in the inner city, she suspects that she wouldn't have been as experimental in her designs had she not been located in her outer suburb.

Respondents typically work across several geographical sites and in multisite workplaces, reflecting the diversity of their work roles and lifestyle arrangements. Most of our respondents work across sites within their suburb, while a handful move between workplaces in the inner city and their outer-suburban home, where they work in a home office or studio. This points to the connection between inner and outer localities and is to some extent representative of changes in broader patterns of work and workplace mobility and flexibility (Gurstein 2001). In some instances it represents the multiple roles that many creative industries workers pursue in order to make a living (Gibson, Murphy, and Freestone 2002). For example, it is typical of musicians and visual artists, at the lower end of the income scale, to have other "day" jobs such as teaching in their area of expertise. This is a common finding among the visual artists and musicians interviewed.

Living in an outer suburb therefore means it is necessary for some respondents to commute to places beyond where they live. Commuting time from Redcliffe to the city by public transport is about an hour. There are no direct train lines to the city but there are direct buses. Interestingly, several respondents who commute, typically on a part-time basis, regard the time on public transport as useful because it allows them time to spend on their creative work. For example, Charlotte, a singer-songwriter who also works two days a week in the inner city in a government job, describes commuting time as "invaluable for songwriting" because it gives her time to "dream and think and read and write."

Similarly, Declan, a graphic designer, found the commuting time useful in preparing for work at his inner-city based architectural company, or for doing his own work:

> Because I commute I get the train [from nearby Sandgate], I don't really like driving that much, so the train at least gives me a bit more time to sketch and draw and get ready for work coming in.

What might, on the surface, appear as a negative aspect of outer-suburban life is in fact experienced as a beneficial or positive aspect of multilocational work. Moreover, the experience of working across two or more geographical sites is indicative of the connections between inner and outer urban areas: As sites of creative industries, inner and urban areas are closely linked.

Locality and Networks

Professional and social networks are identified as critical to the success of the creative economy (Florida 2002). In creative city discourse many of these networks are clustered around inner-city hubs (ibid.). In Redcliffe creative networks also exist, with respondents' networks being diversely represented in the data. Only 7 percent of respondents claim to have no professional networks they considered important, whereas 65 percent have professional networks that span local, national, and international scales. Although much "creative place" thinking assumes that key networks are based on close geographical proximity (Adkins et al. 2007), outer-suburban creative industries network patterns demonstrate that "proximity networks" are only one among a multiscalar complex of professional networks. As Beauregard (1995, 240) argues, "Actors simultaneously have interests at multiple spatial scales; that is, their activities spread out over different geographical fields." For our respondents, technology is a predictably significant factor in creating and sustaining networks beyond the suburb. This is especially evident among multimedia developers, graphic artists, and musicians. Jack, a musician in a band whose members live on the other side of the city from Redcliffe, states that distance is not a problem because of the easy exchange of sound files via the Internet. The band is able to compose songs this way and then rehearse at a location equidistant for each member. Tony, an architect whose practice is in the suburb but who has clients throughout the adjacent region, echoed others' remarks about the facility of technology to create and sustain professional networks. For Tony, efficient local transport routes were also cited as enabling professional connectivity:

> With technology and the transport corridor and what not, I don't find it [lack of local networks] much of an impediment. You just have to manage your time properly.

Our respondents are also active in drawing on and fostering local, proximity-based networks within Redcliffe. John, a graphic designer and multimedia developer, sees locality as creating a unique business edge that provides a good opportunity for networking. John states that the locality can be good for business because the "community feel" is an important leverage point for local businesses: people like to support their local businesses. He also creates and makes use of local services, using the services of local photographers and other allied freelance personnel.

The Redcliffe peninsula has an established artistic community with local networks focused upon the two local galleries that have been operational for many years as hubs of social and professional interaction. Our research found a strong local network of twenty-five-to-thirty-nine-year-old artists whose homes are hubs at which they gather on a weekly basis. The group is active in developing a local creative identity and coordinates events such as a small-scale festival and markets where the emphasis is on locally produced music, art, crafts, and foods. In addition to promoting a local identity, one respondent said that an aim of the festival is "to try and put on an event that's something that we'd want to go to, and hopefully that attracts those people that we want to meet." The importance of "network sociality" that combines both work and play is a recognized feature of creative industries workers (Wittel 2001).

Understanding the professional networks of outer-suburban creative industries workers has clear implications for "creative place" policies. Not all productive creative industries professional networks are proximity-based clusters, and in less dense areas such as outer suburbs,

proximity-based clusters may not at all resemble those found in inner cities. That does not mean that creative clusters do not exist in the suburbs. The multiscalar networks of our interview respondents highlight the complex geographies of the creative industries, and suburbia more generally.

CONCLUSION

This preliminary research into creative workers in a Brisbane outer suburb indicates that, as Gibson and Brennan-Horley found in Sydney, there is much evidence of current and potential creative industries activity in outer suburbia. The research so far has shown the extent to which creative industries workers in outer suburban areas work across multilocational sites, and the importance they attach to their locale as a place removed from the distractions and the busyness of inner-urban life. They are well networked in general, facilitated by technology and efficient road systems, and there is some evidence of creative clustering.

Soja's (1996) assertion that space is physical, imagined, and experienced is important when considering the disjuncture between the imagined space of creative city policy and writings, and the experiential and material space of outer suburban areas. In the case of the Australian "creative place" thinking, the imagined space of policies and planners, and the empirical and experiential spaces of outer suburbia do not currently marry up: Studies might give evidence of creative industries activity and growth in the outer suburbs, but the imagined geography of the suburbs continues to construct the suburbs as banal, unproductive, uncreative, and outside of the glowing center of the inner-city creative core.

The capacity to link creative enterprises to the new suburbanism presents itself as a major opportunity in the development of an innovation culture and economy able to generate wealth and job opportunities through new enterprise formation in growing, high-value-added sectors. However, there are a series of recurring conceptual issues that need to be critically addressed before such a policy strategy can be pursued. It is easy to think about cities in terms of simplistic inner-outer binary; it's also easy to make policy in this way. However, it is time to move beyond this traditional, nostalgic binary. Failing to do so means perhaps disregarding an entire component of the creative industries in Australia, and misunderstanding the nature of those industries. It may also result in creating an inner-outer socioeconomic binary geography in Australian cities if creative place funding and policy energy is predominantly focused on inner cores. To avoid these outcomes, it is time to rethink the suburbs, bringing together their imagined geographies with their experiential and material aspects.

REFERENCES

Adkins, B., M. Foth, J. Summerville, and P. Higgs. 2007. Ecologies of innovation: Symbolic aspects of cross-organisational linkages in the design sector in an Australian inner-city area. *American Behavioural Scientist* 50:922–34.

Australian Bureau of Statistics. 2001. Brisbane statistical portrait 2001. http://www.brisbane.qld.gov.au (accessed March 3, 2005).

———. 2006. 2006 census. http://www.abs.gov.au/websitedbs/ d3310114.nsf/home/Census+data (accessed January 12, 2009).

Bathelt, H., A. Malmberg, and P. Maskell. 2004. Clusters and knowledge: Local buzz, global pipelines and the process of knowledge creation. *Progress in Human Geography* 28:31–56.

Beauregard, R. 1995. Theorizing the global-local connection. In *World cities in a world system*, ed. P. Knoz and P. Taylor, 232–48. Cambridge: Cambridge University Press.

Cairncross, F. 1998. *The death of distance: How the communications revolution will change our lives.* Cambridge, MA: Harvard University Press.

Castells, M. 1989. *The informational city: Information technology, economic restructuring and the urban-regional process.* London: Blackwell.

Danaher, G. 2007. The region as performance space: A distinctive take on the creative industries. *Studies in Learning Evaluation, Innovation and Development* 4:11–19.

Department of Communications, Information Technology and the Arts. 2002. *The creative industries cluster study, vol. 1.* Canberra: Commonwealth of Australia.

Department for Culture, Media and Sport, UK. 1998. *Creative industries mapping document.* London: Author. http://www.culture. gov.uk/reference_library/publications/4740.aspx (accessed April 2, 2009).

Drake, G. 2003. "This place gives me space": Place and creativity in the creative industries. *Geoforum* 34:511–24.

Florida, R. 2002. *The rise of the creative class.* Cambridge, MA: Basic.

———. 2004. Creative class war. *Washington Monthly*, January 15. http://geography.berkeley.edu/ProgramCourses/CoursePagesFA 2004/Geog110/CreativeClassWar.Florida.pdf (accessed March 20, 2009).

———. 2005. *Cities and the creative class.* London: Routledge.

Gertler, M. 2004. *Creative cities: What are they for, how do they work, and how do we build them?* Canadian Research Policy Networks, Family Network, Paper F/48. http://www.cprn.org/doc.cfm?doc= 1083&l=en (accessed March 20, 2009).

Gibson, C., and C. Brennan-Horley. 2006. Goodbye pram city: Beyond inner/outer zone binaries in creative city research. *Urban Policy and Research* 24:455–71.

Gibson, C., and J. Connell. 2003. Bongo fury: Tourism, music and cultural economy at Byron Bay, Australia. *Tidjschrift voor Economische en Sociale Geografie* 94:164–87.

Gibson, C., and L. Kong. 2005. Cultural economy: A critical review. *Progress in Human Geography* 29:541–61.

Gibson, C., P. Murphy, and R. Freestone. 2002. Employment and socio-spatial relations in Australia's cultural economy. *Australian Geographer* 33:173–89.

Gipps, P., J. Brotchie, D. Henscher, P. Newton, and K. O'Connor. 1996. *Journey to work: Employment and the changing structure of Australian cities.* Melbourne: Australian Housing and Urban Research Institute.

Gleeson, B. 2002. Australia's suburbs: Aspiration and exclusion. *Urban Policy and Research* 20:229–32.

Gurstein, P. 2001. *Wired to the world, chained to the home: Telework in daily life.* Vancouver: University of British Columbia Press.

Hartley, J., ed. 2005. *Creative industries.* London: Blackwell.

Helbrecht, I. 1998. The creative metropolis: Services, symbols, and space. *International Journal of Architectural Theory* 3. http://www .theo.tu-cottbus.de/wolke/X-positionen/Helbrecht/helbrecht.html (accessed March 20, 2009).

Higgs, P., S. Cunningham, and J. Pagan. 2007. *Australia's creative economy: Basic evidence on size, growth, income and employment.* Technical report. Brisbane, Australia: ARC Center of Excellence for Creative Industries & Innovation. http://eprints.qut.edu.au/8241 (accessed March 20, 2009).

Horne, D. 1964. *The lucky country: Australia in the sixties.* Ringwood, Australia: Penguin.

Hutton, T. 2006. Spatiality, built form, and creative industry development in the inner city. *Environment and Planning A* 38:1819–1841.

Landry, C. 2000. *The creative city: A toolkit for urban innovators.* London: Earthscan.

———. 2003. *Rethinking Adelaide: Capturing imagination.* Adelaide: Department of the Premier and Cabinet.

Kanigel, R. 1997. *The one best way: Frederick Winslow Taylor and the enigma of efficiency.* New York: Viking.

Kinnane, G. 1998. Shopping at last! Fiction and the anti-suburban tradition. In *Writing the everyday: Australian literature and the limits of suburbia*, ed. A. McCann, 41–55. Brisbane: University of Queensland Press.

Kong, L. 2005. The sociality of cultural industries: Hong Kong's cultural policy and film industry. *International Journal of Cultural Policy* 11:61–76.

Kotkin, J. 2005. *The new suburbanism: A realist's guide to the American future.* Costa Mesa, CA: The Planning Center. http://www.joelkotkin.com/Urban_Affairs/The percent20 Newpercent20Suburbanism.pdf (accessed March 20, 2009).

LeFebvre, H. 1992. *The production of space*, trans. D. Nicholson-Smith. Oxford: Blackwell.

Lubart, T., and R. Sternberg. 1998. Creativity across time and place: Life span and cross-cultural perspectives. *High Ability Studies* 9:59–73.

Malanga, S. 2004. The curse of the creative class. *City* 41(1). http://www.city-journal.org/html/14_1_the_curse.html (accessed March 20, 2009).

McCann, A. 1998. Introduction: Subtopia, or the problem of suburbia. In *Writing the everyday: Australian literature and the limits of suburbia*, ed. A. McCann, vii–x. Brisbane: University of Queensland Press.

Menin, S., and F. Samuel. 2003. *Nature and space: Aalto and Le Corbusier.* London: Routledge.

Miles, M. 2005. Interruptions: Testing the rhetoric of culturally-led urban development. *Urban Studies* 42:889–911.

National Economics. 2002. *The state of the regions.* Melbourne: National Economics and the Australian Local Government Association.

National Institute of Economic and Industry Research, Australia. 2004. *Melbourne creativity.* Report to the Innovation Economy Advisory Board, Government of Victoria. Melbourne: National Institute of Economic and Industry Research.

Oakley, K. 2004. Not so cool Britannia: The role of the creative industries in economic development. *International Journal of Cultural Studies* 7:67–77.

O'Connor, K. 1999. Misunderstanding modern suburban development. *People and Place* 7:21–25.

O'Connor, K., and R. Stimson. 2004. Population and business in metropolitan and non-metropolitan Australia: The experience of the past decade. *People and Place* 5(2). http://elecpress.monash.edu.au/ pnp/free/pnpv5n2/oconkath.htm (accessed March 20, 2009).

Porter, M. 2000. Location, competition, and economic development: Local clusters in a global economy. *Economic Development Quarterly* 14:15–34.

Pratt, A., R. Gill, and V. Spelthann. 2007. Work and the city in the e-society: A critical investigation of the socio-spatially situated character of economic production in the digital content industries in the UK. *Information, Communication and Society* 10:922–42.

Queensland State Government, Industry Development. 2008. *About creative industries*. Brisbane, Australia: Author. http://www.industry .qld.gov.au/dsdweb/v4/apps/web/content.cfm?id=6738 (accessed March 10, 2009).

Randolph, B. 2004. The changing Australian city: New patterns, new policies and new research needs. *Urban Policy and Research* 22:481–93.

Ross, A. 2006–7. Nice work if you can get it: The mercurial career of creative industries policy. *Work Organisation, Labour, & Globalisation* 1:1–19.

Salt, B. 2006. *The big picture: Life, work and relationships in the 21st century*. Melbourne: Hardie Grant.

Scott, A. 1999. The cultural economy: Geography and the creative field. *Media, Culture and Society* 21:807–17.

Soja, E. 1996. *Thirdspace: Journeys to Los Angeles and other real-and-imagined places*. Cambridge, MA: Blackwell.

Stimson, R. 1998. Dynamics of Brisbane's inner city suburbs. *Australian Planner* 35:205–14.

Turner, G. 2008. The cosmopolitan city and its Other: The ethnicising of the Australian suburb. *Inter-Asia Cultural Studies* 9:568–82.

United Nations Conference on Trade and Development. 2008. Secretary-general's high level panel on the creative economy and industries for development. United Nations. http://www.unctad.org/ en/docs/tdxiibpd4_en.pdf (accessed March 10, 2009).

Wittel. A. 2001. Towards a network sociality. *Theory Culture & Society* 18:51–76.

Zukin, S. 1995. *The culture of cities*. Cambridge, MA: Blackwell.

GIS, Ethnography, and Cultural Research: Putting Maps Back into Ethnographic Mapping

Chris Brennan-Horley

Australian Centre for Cultural Environmental Research, School of Earth and Environmental Sciences, University of Wollongong, Wollongong, New South Wales, Australia

Susan Luckman

School of Communication and Hawke Research Institute, University of South Australia, Adelaide, South Australia, Australia

Chris Gibson

Australian Centre for Cultural Environmental Research, School of Earth and Environmental Sciences, University of Wollongong, Wollongong, New South Wales, Australia

Julie Willoughby-Smith

School of Communication, University of South Australia, Adelaide, South Australia, Australia

This article discusses how geographic information system (GIS) technologies were used to enhance ethnographic methodologies within a cultural research project, *Creative Tropical City: Mapping Darwin's Creative Industries*. It shows how mapping technologies can broaden the scope of data available via interview practices and produce innovative ways of communicating research results to stakeholder communities. A key component of the interview process was a "mental mapping" exercise whereby interviewees drew sketches, revealing important sites and linkages between people and places. A GIS linked responses to real-world locations, collating and displaying them in meaningful ways. Responses uncovered Darwin's unique geography of creative inspiration—a geography that preferences Darwin's natural environment over sites of urban creative milieu.

Geographic information systems (GIS) are increasingly popular technologies within both the academy and the public sphere—from origins in applied research in physical environmental sciences, to the incorporation of digital mapping technologies into mobile phones and car navigation devices. There is growing interest in the use of GIS for cultural research, pioneered within human geography, but now increasingly applied within media and communication studies and cultural studies (e.g., Elwood 2006a, 2006b; LeGates 2005; Nash Parker and Asencio 2008; Steinberg and Steinberg 2006). Topics have included use of GIS to better understand community aspirations and responses to neighborhood renewal programs (Elwood and Leitner 1998); GIS as a participatory research tool (Vajjhala 2005); mapping fear and misperceptions of urban space (Matei, Ball-Rokeach, and Qiu 2001); and spatial histories of cinema attendance (Allen 2008; Klenotic 2008). This article extends this burgeoning interdisciplinary field, by discussing how GIS technologies are being used to enhance the ethnographic methodologies—specifically interviews—within a research project principally about documenting the presence and character of the creative industries in a small, remote Australian city.

Our focus on incorporation of GIS into a creative city research project's ethnographic methods comes at a time when "mapping" is an increasingly popular phrase used in cultural studies, media, and creative industries research. Often, though, the phrase is used metaphorically rather than literally, referring instead to a synthetic discussion or overview of a particular aspect of media culture (e.g., Montgomery 2000; Primorac 2004). In such studies, the metaphorical application of "mapping" may well be appropriate, but no actual maps in the literal sense are produced or analyzed. In parallel to this, where interviewing within creative industry research on the role of space and place has been the main focus (e.g., Drake 2003), such interviews have similarly not yielded actual mappable data. The embedding of mental mapping exercises in qualitative interviews brings together methodologies from cultural studies and cultural geography to connect the physical world with the imagined one, bringing the visual representation of space and meaning back into humanities and social sciences–based "mapping" projects.

References to "mapping" as a methodology have been particularly frequent of late in creative industries research. Where geographical methods have indeed been used, orthodox techniques for mapping the creative economy have attempted to categorise the size, economic significance, and growth patterns of creative industries, with cities usually the spatial unit of analysis. Commonly used methods include the analysis of intrafirm and industry structures; quantifications of creative workers as components of local labor markets (e.g., through census data analysis on employment across contiguous spatial units); and mapping firm locations across cities or regions with accompanying analysis of "cluster effects" (see Gibson and Brennan-Horley 2006; Gibson, Murphy, and Freestone 2002; Markusen, Wassall, and DeNetale 2008; Pratt 1997; Scott 2000; Throsby 2001; Watson 2008). Despite the shortcomings and applicability of statistical methods for creative city research (cf. Pratt 2004), these techniques are regularly used in cities or regions containing substantial populations (usually greater than 500,000 inhabitants), a situation not found in our case-study location, Darwin, which has a population of 75,000 (ABS 2006).

Darwin is a small place with a creative workforce of approximately 1,800 individuals (Brennan-Horley and Gibson 2009). Although census data mapping was undertaken using statistics on this creative labor force (see Gibson and Brennan-Horley 2007), this was of limited utility—given lack of critical mass, limited resulting sample sizes in individual workforce categories, and Darwin's own small geographical extent. Even more so than elsewhere, people in Darwin's creative economy (particularly independent cultural producers) rely heavily on unpaid, volunteer, and *pro-am* (amateurs operating at a professional quality but not getting paid for it) patterns of labor. Creative employment is affected by the short-term life cycle of most projects, with practitioners regularly moving between the informal and formal sectors of the creative economy (Shorthose 2004). The national census and the Australian business register system, both premised on capturing stable, full-time employment and business activities, routinely underestimate the extent of informal creative pursuits, and when applied to Darwin, such data proved grossly inadequate. There was an unknown proportion of creative workers—whose activities were economically marginal, or were undertaken as a secondary form of employment—who remained absent from official statistics for Darwin (cf. Gibson, Murphy, and Freestone 2002).

Such shortcomings are not unique to Darwin (though they were particularly magnified there), hence the shift within creative city research toward the microprocesses that govern the nature of work in the creative industries and its spatial manifestation (Gibson 2003; Brennan-Horley 2007; Luckman 2008; Luckman, Gibson, and Lea 2009; McRobbie 2002; Shorthose 2004; Pratt 2004; Neff 2005; Neff et al. 2005). These microprocesses can only be uncovered through detailed ethnographic research. For these reasons, interviewing creative industry practitioners was central to our research from the very beginning. Yet, given Darwin's small geographical size and particular reliance on informal sector creative work, interviews took on a heightened importance. Interviews were by necessity the main avenue to generate overall data on the creative industries in Darwin. Furthermore, over the course of the project, interviews became a possible site of gathering spatially tagged data that could be used in GIS analysis—enabling mapping work to be done from within a methodology focused principally on documenting informal and hidden patterns of work. What started as a separate means of obtaining relevant qualitative data on work in the creative industries became a synthetic method for generating new geographically specific data on questions of "where" creativity exists in the city, when orthodox quantitative methods failed. Interviews thus became a much richer, innovative avenue of inquiry.

Here, we focus on this story of how GIS technology was used to prompt, gather, and analyze interview data on creative industries in a place. Specifically, this article focuses on two methodological advantages of GIS: how mapping technologies can be used to broaden the scope of data available via interview practices, and how they produce innovative new ways of effectively communicating research results back to stakeholder communities. To our knowledge, nowhere else have such attempts been made. Our ensuing discussion builds on these advances, noting the relevance and possibilities of GIS for university research in a new phase of media literacy. Digital map databases provide an accessible and meaningful interactive interface that can be used by researchers for the purpose of information dissemination and dialog. Maps of place provide a readily identifiable and user-friendly point of entry into university-generated research, and they additionally open up the methodologies we employ to the critical scrutiny of a wider audience of stakeholders.

PROJECT BACKGROUND

The Creative Tropical City project sought to explore the idea of Darwin, a place in Australia's tropical-savannah north, as a "creative city." Funded through the Australian government's Australian Research Council Linkage Project scheme, the project was jointly supported by Darwin City Council, the Northern Territory Tourism Commission, and the Northern Territory Government Arts and Museum Division of the Office of the Chief Minister, who were each interested in pursuing new policies to enhance Darwin's creative industries, liveability, and attractiveness to new migrants. Specifically, the research project's three stated aims were:

- To determine the nature, extent, and change over time of the creative industries in Darwin;
- To interrogate the applicability of national and international creative industry policy frameworks to Darwin; and
- To identify opportunities for transformation in the creative industries in Darwin.

Darwin is a small city located on the northwestern tip of Australia's Northern Territory, with a large Aboriginal population and a highly transient professional workforce. It is one of Australia's most remote capital cities, nearly 1,500 km from the nearest substantial town of Alice Springs (a town of only 25,000 inhabitants) and 3,000 km from its nearest neighboring capital city, Adelaide in South Australia. Darwin thus has a special place within Australia's cultural imagination—as a strategic northern military outpost on the frontier, a tourist gateway to the "Top End" wilderness, and a focal point for postcolonial struggles over land rights and mineral extraction (Jull 1991).

Contrasting with this national sense of remoteness, Darwin is spatially proximate to Southeast Asia, only 700 km away from East Timor and in close proximity to the populous centers of Jakarta and Singapore. This closeness is reflected in the city's highly multicultural population, which includes a number of well-established Southeast Asian communities (ABS 2006). Darwin is situated on a peninsula, with its internal spatial layout shaped by this position, as well as by a number of catastrophic events in its tumultuous, short European history. First laid out for development in the late 1890s, the city has been destroyed four times: twice by cyclone, by fire, and by bombing raids during World War II (as recently depicted in Baz Lurhmann's epic motion picture, *Australia*). Cy-

clone Tracy, which hit on Christmas Eve 1974, demolished the city, resulting in a new replacement city fashioned with architectural styles and spatial layouts characteristic of post-1970 planning schemes: low-density, cyclone-proofed buildings, located in predominantly suburban residential neighborhoods, connected by wide ring roads, with high dependence on air conditioning somewhat imprisoning inhabitants from their tropical surrounds. The city layout is further split in two by its international airport (which is disproportionately large because of its important strategic air force base, deliberately built to handle large planes and volumes flying to and from the Asia-Pacific). In effect, Darwin's comparatively affluent northern suburbs are separated by the airport from the city's small central business district (CBD) and port/warehouse districts located to the south. Lower income and (once informal) fringe Aboriginal settlements ring the city. This is the geographical context underlying the present analysis of the geography of creativity in Darwin.

Because Darwin has such a radically different set of geographic circumstances, it diverges from the typical places in creative city literature—often deindustrializing or prominent knowledge-economy cities in the Northern Hemisphere such as Manchester, Liverpool, Barcelona, Amsterdam, New York, Boston, Los Angeles, and San Francisco (e.g., Scott 2000, 2004; Markusen, Wassall, and DeNetale, 2008)—or their Southern Hemisphere counterparts (e.g., Gibson and Brennan-Horley 2006; Cunningham, Hearn, Cox, Ninan, and Keane 2003). Unlike these well-known creative cities, Darwin has no brownfields sites ripe for reinvention as "creative precincts," and apart from Mitchell Street, the center of the city's backpacker scene, it does not have European-style concentrations of cafés and bars. Darwin simply lacks the types of urban environments that other studies have suggested are important for the generation of creative milieu (Scott 2000; Drake 2003). Despite this, Darwin still has visible signs of an active creative economy. It hosts many festivals during the dry season, has a vibrant market culture unlike that of any other major Australian capital, it is the center of international trade in indigenous art from surrounding top-end communities, and has audible alternative music scenes. The disjuncture between archetypal "creative cities" and Darwin's peculiar mix of remoteness, smallness, internal diversity, and environmental conditions presented an opportunity to try new methods of understanding the spatial dimensions of creative industries within a city. We sought different methods to map where creativity was present in Darwin, and how it related to the city's unique urban form.

To this end, between April 2007 and February 2008, we conducted interviews with creative industry practitioners, generating over 600,000 transcribed words, as well as almost 100 maps and other data. In total, 98 semistructured interviews were conducted with creative practitioners

TABLE 1
Types of creative workers interviewed, 2007–8 (number of interviews)

Actor (5)	Dancer (2)	Multimedia designer (3)
Actor/director (1)	Designer (1)	Musician (7)
Advertiser (1)	Director (1)	Performing artist (4)
Architect (7)	Educator (4)	Photographer (6)
Artist (5)	Entertainment reporter (1)	Playwright (1)
Artistic director (5)	Festival and event manager (5)	Printmaker (1)
Arts administrator (22)	Filmmaker (2)	Promoter (3)
Arts retailer (2)	Goldsmith/jeweler (2)	Publishing (3)
Author (3)	Graphic designer (1)	Retail (2)
Building designer (1)	Horticulturalist and landscape consultant (1)	Sound artist (1)
Building surveyor (1)	Leatherworker (1)	Sound engineer (3)
Choreographer (3)	Manager (2)	Tattooist (1)
Cinematographer (2)	Manufacturing and sales (3)	Venue manager (1)
Clothing, bags, and jewelery designer/manufacturer (2)	Marketing (2)	Video producer (1)
Community arts worker (3)	Media presenter, journalist, producer (7)	Visual artist (15)
Curator (1)		

working and/or living in the Darwin area (see table 1). Sampling methods included listings in the Darwin Yellow Pages, recommendations, expressions of interest from potential participants themselves, fieldwork, Web searches, and local media. Two interview schedules—one for those who work in a creative capacity as their primary employment and another for those not employed in creative industries as their primary source of income—were employed. Through analysis of the data generated through the interviews, we were able to better able to identify the "soft infrastructure" required for creative industries development; how key institutions and actors in the creative industries related to each other; how workers in the creative industries perceived Darwin's potential as a creative place; the limitations and barriers currently encountered by creative industries in Darwin; and how the creative industries in Darwin contributed to current economic growth, employment, and community development—knowledge essential to successfully delivering on the aims of the project.

But still, there remained a need to address questions of *where* creativity resided in the urban environment: the day-to-day interactions, movements, and work patterns of creative individuals matter because it is their movements between the various sites and spaces of a city that facilitate creative practices and activity (Scott 2000). Furthermore, what were the qualities of these sites that influenced and enabled creativity? These sorts of questions could not be answered with employment statistics (that were problematically inaccurate for Darwin anyway), but rather required the construction of maps of creative Darwin, built up in a qualitative fashion through the responses of practitioners to interview questions. We subsequently explored the use of a mental mapping exercise during the interview process.

MENTAL MAPPING THE CREATIVE CITY

A mental map is an individual's cognitive representation of place (Taun 1975). Similar terms, including sketch maps and cognitive maps, cover similar conceptual ground, but essentially they are all used to elicit the importance a respondent gives to particular spaces, sites, and nodes in networks, the strength of the relationships between these, and how these factors combine to create one's orientation, goals, or mental geography of place (Lynch 1960; Matei, Ball-Rokeach, and Qiu 2001). Mental maps have been deployed as a research tool since the 1960s, emerging from the field of behavioral geography, and historically were primarily carried out on paper, with an informant sketching a freehand map of an area in question.

Being hand drawn, mental maps have not in the past been too concerned with geographical accuracy. Distortion makes cartographic comparisons between informant responses difficult and little work has been done in computer-aided mental map comparison. However, recent advances in GIS software are permitting measurement of distortion levels between localized mental maps (see Peake and Moore 2004) and are allowing for digital recording and analysis of line segments via the use of tablet personal computers (PCs) and digital pens (see Huynh and Doherty 2007). For the Darwin project we were concerned with building up a data set based on informants' responses to particular areas of their city, followed up with questions about the characteristics of those sites that made them important to their creative activities (rather than uncovering information about informants' spatial cognition, as has been the focus of much previous work on mental mapping). What was needed for our project was a method that would remove the distortion to allow for quicker comparison between responses. Drawing on similar studies that have attempted to meld GIS with mental mapping (see Matei, Ball-Rokeach, and Qiu 2001; Vallhaja 2004), we enlisted the use of a paper base map made up of Darwin's statistical local area (SLA) boundaries and a road network (see figure 1). This meant that every informant's base map was identical. Any sketching that took place upon them could be easily digitized and georeferenced (assigning a coordinate system of location in virtual space) from the base map and stored for further analysis with the GIS.

The mental mapping exercise was woven into the interview process, with spatial questions asked at particular junctures. Respondents marked their base map using different colored pens for each question while expanding on their answers verbally.[1] Within the framework of a semistructured, recorded interview, we found it notable that the mental mapping exercise activated a different attitude to the interview situation on the part of the participants than a more straightforward and expected question-and-answer format. Drawing on the map prompted people to think visually about their connection to Darwin and the relationship between places, both within the city and beyond it. Putting different colored pens into their hands and asking them to "draw," forced participants to express themselves in a manner beyond the usual publicly performed self of typical interview respondents. Also, as a nonlinguistic and highly visual means of representation, mapping can be challenging to those comfortable in their verbal selves, but less comfortable visually. Drawing is another way of expressing the world, and while challenging for some, it also aided in overcoming the inherent English-language proficiency bias of the interview mode.

UNCOVERING DARWIN'S GEOGRAPHY OF CREATIVE INSPIRATION

A key function of GIS is its ability to display multiple layers concurrently, known as an overlay. This is a simple yet powerful way of revealing spatial relationships between layers of information. Once mental maps were collected from the interviews, they were each georeferenced and stored in the same coordinate system (and hence would "line up" on top of one other) using ArcGIS 9.2 software. Responses to each spatial question were then digitized and transferred to the GIS. Each response effectively acted as its own layer in the GIS. Data sets made up of the responses to each question could then be collated, compared, and visualized using different extensions available in ArcGIS, including Spatial Analyst, 3D Analyst, and ArcScene.

Figure 2 is a composite of all responses to the question, "Where do you go for creative inspiration?" The aim of this question was to get a sense of any underlying patterns for this important but hitherto unseen aspect of creative work. In total, seventy-seven responses were given to this question and have been compiled to make the map shown in figure 2. Darker shades denote areas where greater numbers of respondents agreed that that space was inspiring for their creative activities. There was a distinct concentration of responses situated around particular shorelines, most notably around Nightcliff and Fannie Bay, places that approximately 26 percent of responses agreed upon. A degree of spatial correlation exists between the existence of natural features, such as parks, gardens, and beaches, and spaces of creative inspiration. This was further elaborated in the interview transcripts, with one musician responding:

> "Yeah I find being outdoors because, things come to you a bit more, you can relax more, I find I relax more when I'm outdoors and when I'm near water . . . I can start the creative process by sitting down with my note pad and writing stuff out." (pers. comm.)

Other respondents agreed, marking on their maps the shorelines and beaches of popular locations such as Nightcliff and Fannie Bay, while elaborating further about their qualities. For one visual artist (pers. comm.): "Well the beach I know that always makes me feel at peace"; similarly, for a theatre director (pers. comm.):

> "I'm not a visual artist so a lot of what I do goes on in my own head so I guess just places like the beach where you can go and think and have a bit of space."

The soothing and relaxing aspect of Darwin's natural environment was a recurring theme.

In contrast, there was talk of other more "urban" spaces around Darwin that were inspiring. For one local film director (pers. comm.):

> "Well it's not so much the physical or geographic things—although I have to say the Nightcliff foreshore I've used quite a lot, yeah definitely that's been a source of inspiration, East Point similarly and I think to some extent the Fannie Bay foreshore as well—but then you could say at the same time Winnellie in the kind of culture of the warehouses and the wrecking yards and all that sort or stuff can be quite inspiring as well . . . the industrial side of Darwin."

Here again is a reference to the striking features of western foreshore areas, but also to other more industrialized spaces associated with Darwin's working harbor, thereby indicating that creative inspiration in Darwin is not entirely confined to its natural features.

Not all respondents spoke about specific places—for some, creative inspiration came from Darwin as a whole. Seven percent of respondents felt that all of Darwin was inspiring, as opposed to one or two particular spaces. Their responses on the map consisted of a shape that encapsulated the whole city (figure 2). Although such responses could be viewed as being not spatially "accurate" or too broad, they were just as valid—they simply reflect a

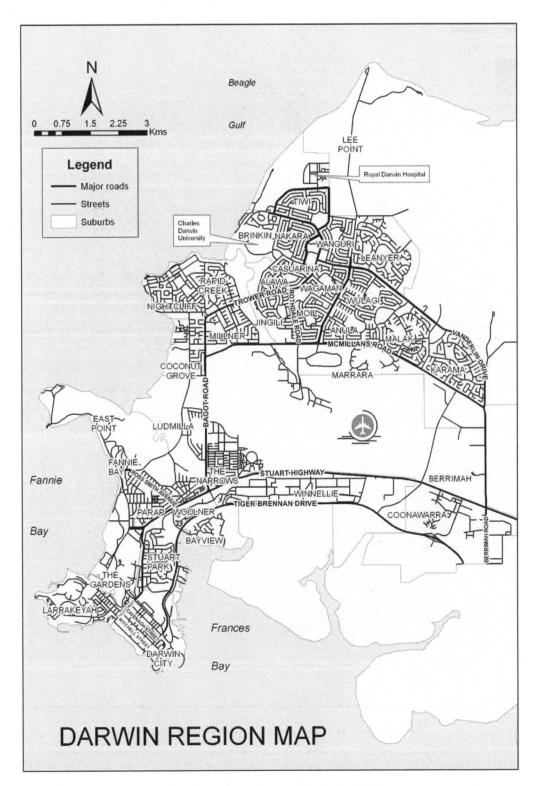

FIG. 1. Example of a blank mental map.

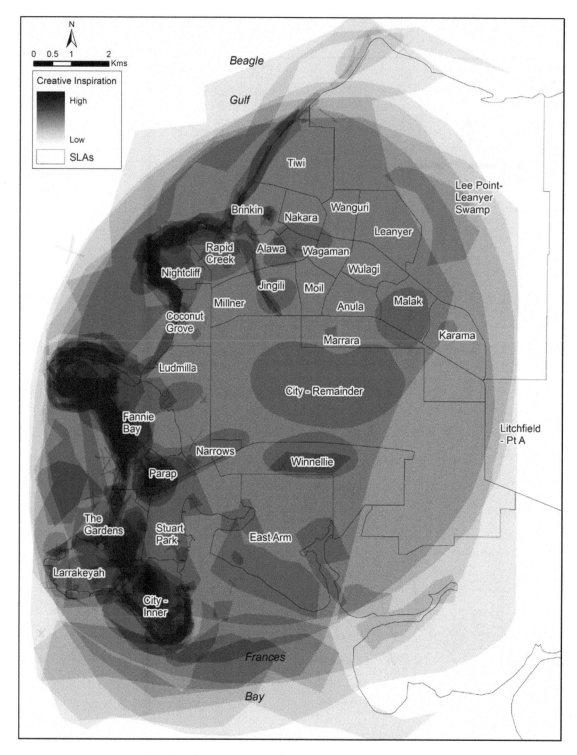

FIG. 2. Composite map of creative inspiration responses.

geographical imagination and sense of place enacted over a larger scale. There was a temptation in the analysis phase to normalize the data, treating these responses that were not spatially "specific" as outliers, but doing so would be to effectively ignore that creatively inspiring places can operate at different scales, manifest in different locations, and, most importantly, are highly personal in nature. For some creative practitioners, creative inspiration is about the natural environment, a feeling of closeness and having a degree of access to such spaces. For others, it was more about the spaces in which they worked or operated, or was about what Darwin as a whole city offered. Taking this last point a step further, one respondent noted on the map "people give me inspiration" and then circled this with a blue pen—a spatially abstract but nonetheless valid response.

Another key feature of GIS is its ability to represent data sets in a variety of ways. Figure 3 is an example of the previous data (in figure 2) extruded in three dimensions—sites of repeated mention in response to an interview question are represented as mountain peaks, and rarely mentioned places as low-lying plains. For example, a ridge effect running along the western coast is more starkly evident when visualized in this manner, as are other concentrations of inspiration in the suburbs of Parap, the Botanic Gardens, and parts of the city—the CBD. Additionally, three-dimensional (3D) animation schemes can be set up as an interactive tool, whereby individuals using a computer mouse are able to effectively "fly through" the 3D map of the creative city, shifting the camera view along various tracks to better observe the range of peaks of creative inspiration. Although it is beyond the scope of this article, this visualization technique is being used as a means to compare against other data sources generated by our interviews—for example, further analysis in the project examines how this particular geography of inspiration intersects with other geographies of creative epicenters, zones of recreation, types of work sites, and relationships between work and home. A range of iconic and mundane elements in people's creative lives can be graphically represented together in this manner as a 3D visual morphology.

The capabilities of GIS databases extends further into hyperlinking geographical locations back to qualitative interview responses in written or audio form, or to other visual data sources such as digital photographs. This method allows for a variety of data sources to be incorporated into the interactive map, adding a further level of detail to important sites, and reinforcing why respondents felt they are a vital part of the creative city. The research process culminated in a high-profile, public exhibition that launched results of the Creative Tropical City project in Darwin's main city library in February 2009. The maps were made available for public viewing, as well as being archived on the Internet through a blog site (http://creativetropicalcity.blogspot.com), for stakeholders to register their responses to the map data and to the wider project. It was envisaged that this map interface would act as a feedback mechanism and conduit for further interactions between creative industry practitioners and those sectors of the Northern Territory (NT) government responsible for pushing creative industry policy forward after the completion of the CTC project.

DISCUSSION: THE VALUE OF USER-GENERATED MAPS IN QUALITATIVE CREATIVE INDUSTRIES INTERVIEWS

Cultural geographers have unpacked the meanings and discursive power that underlie cartography and map-making (Harley 1989; Rose-Redwood 2006). Maps have the ability to concretize realities in the eyes of the observer, giving the cartographer the power to choose particular realities and (re)create them. Through the maps presented here and others created for the project we can visually make known to the wider community that Darwin is a creative city, cognizant that these maps are among the many ways in which the city could be imagined spatially. Policymakers and other stakeholders can view sites that matter to Darwin's creative community.

If the perceptual problem with qualitative research is that it cannot get the message across, so to speak, as clearly as a quantitative analysis can, then this method offers a way to redress the balance. For example, one clear outcome of the mental mapping exercise in the Creative Tropical City project has been to visually communicate the importance of Darwin's parklands to the creative practice of its citizens. As noted on the front page of the *Weekend Australian* in June 2008, Darwin is under land pressure as the population grows, though the city itself has limited space, constrained as it is physically by the Timor Sea and mangrove swamps (Toohey 2008, 1). As a result of this pressure, high-rise apartment blocks on a scale previously unimaginable in the decades after Cyclone Tracy are now seen as the answer to Darwin's prayers. Our research, on the other hand, suggests that a commitment to maintaining its unique natural assets (parklands and beachfronts) for public access should be as central to its creative industries policy as more immediately obvious infrastructure such as shopfronts and training. What enabled us to strongly convey this point to policymakers was the scanning into GIS of this data.

Additionally, digital map databases provided an accessible and meaningful interactive interface that can be used by researchers for the purpose of research information dissemination and dialogue. Maps of place provide a readily identifiable and user-friendly point of entry into university-generated research. This project used mapped outputs as an exhibition centerpiece, showcasing to the Darwin community the various data outputs generated by this project: interview quotes; statistical data; musical background audio track; photographs; and 3D "fly-through" maps of "creative Darwin." As work in the area of participatory GIS (PGIS) illustrates (see Elwood and Leimer 1998, Vajjhalla 2005, Rambaldi and Callosa-Tarr 2000), the use of GIS and other mapping tools as a database for cultural research has the potential to improve the visual, spatial, synchronic, and diachronic analytical tools

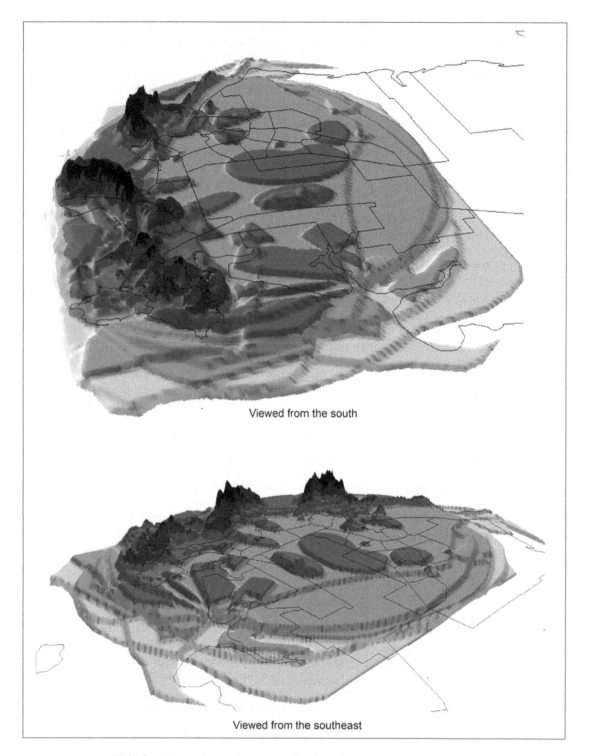

Viewed from the south

Viewed from the southeast

FIG. 3. Three-dimensional visualization of creative inspiration data.

available to cultural researchers, and also open up the methodologies we employ to a wider audience of stakeholders, especially the research subjects themselves.

Moreover, from a methodological perspective, the use of mental maps alongside traditional interview practices acts as a triangulation device, grounding comments in actual space and potentially prompting responses that go beyond the conventional wisdom on creative locales in the city. Asking questions about the actual use of space and getting respondents to represent it on a familiar map connects the performative practices of interview response to the quotidian reality of everyday creative practice. Questions asking respondents to first locate their places of operation, of travel to and from work, and of residence, before moving onto the imagined spaces of creative inspiration, establishes a scenario for interviewees to provide more grounded responses to the question: "Where is creative Darwin?" This process counteracts the tendency of respondents to revert back to conventional notions of creative spaces—the "usual suspects" of branding, policy, and place marketing—thus allowing researchers to drill down into the less obvious but nonetheless significant spaces. It also counters the tendency in qualitative interviews for respondents to overlook the taken-for-granted aspects of their creative practice (Bryman 2008).

A key example was the finding that the largely industrial suburb of Winnellie was a creative hotspot, despite its non-CBD modern warehouses. With the mental maps, questions about people's everyday movements gave rise to responses such as:

> The other area that we travel to that I've just remembered and will mark with the blue pen is Winnellie and I think Winnellie has got a heap of potential . . . we've got an arts space there which is storage, and also it's providing accommodation for arts organisations. I think there's scope there for more to happen and Winnellie is just quite an interesting area. I can see it in the future as perhaps hosting more studios and that sort of thing, although people say it's out of the way etc., industrial, but often it's in the industrial areas where you've got the interesting things that happen. (interview, arts manager, April 2008)

> Yeah this particular part of Winnellie is a pivot, there's several reasons for being here. One is if you're in the centre of town people can't get a park or they have to park a kilometre away and walk and it's just inconvenient even though you've got shops and all the rest of it there, and people and interaction. In Winnellie people can come and park, they don't feel intimidated, it's easy to drop in. Also you're only 5 minutes from anywhere because of the major roads that intersect here; Tiger Brennan Drive is just there so 5 minutes you're in town, 10 minutes you're down at Palmerston, Bagot Road you're straight to the northern suburbs and Stuart Highway so this location here is the pivot of all those roads. (interview, designer, January 2008)

Best accessed by car, it is the antithesis of a creative hub as articulated in conventional creative cities policies, such as those championed by Richard Florida and Charles Landry, who focus on revamped, high-density, urban spaces with a focus on walkability or cycling, and the creative reuse of industrial building stock. As a city prone to repeat disasters (cyclones, World War II bombing), and with no manufacturing history, there is no sprawl of derelict industrial buildings ripe for redevelopment. Unlike northern postindustrial towns in the United Kingdom or United States, Darwin has never had a gentrifiable inner city, and its housing costs are already among the most expensive in Australia (a function of remoteness, supply shortage, and policy incentives for property development as a form of domestic economic vivification). Winnellie's biggest attraction was affordable studio space. In the same way, big-box retailers (such as the recently opened hardware warehouse Bunnings, located near Darwin airport) emerge on maps as unlikely sites in the creativity value chain; in the words of one of our respondents: "You'd be surprised the amount of networking I do at Bunnings" (interview, visual artist, April 2008). For cities that don't fit the standard creativity script (and perhaps also those that do), mental maps are an invaluable triangulation tool for revealing the "unsexy" reality of sustainable creative practice, within the format of the qualitative interview where the temptation may well be to provide the easiest answer, and not necessarily the most accurate (and hence probably messy) one.

Such a shift in research practice responds to challenges facing traditional knowledge generation in the current (new) media age. To quote from a leading Web 2.0 tome, *Wikinomics*:

> The production of knowledge, goods, and services is becoming a collaborative activity in which growing numbers of people can participate. This threatens to displace entrenched interests that have prospered under the protection of various barriers to entry, including the high costs of obtaining the financial, physical, and human capital necessary to compete. Companies accustomed to comfortably directing marketplace activities must contend with new and unfamiliar sources of competition, including the self-organized masses, just as people in elite positions (whether journalists, professors, pundits, or politicians) must now work harder to justify their exalted status. (Tapscott 2007, 16–17)

Taking this even further, publicly available map-based methodologies provide researchers with potential ways in which user-generated content can become a part of the research process. To quote Tapscott (2007) again:

> The time to address peer production is now. Barriers to entry are vanishing and the trade-offs that individuals make when deciding to contribute to projects and organizations are changing, creating opportunities to dramatically reconfigure the way we produce and exchange information, knowledge, and culture. (ibid., 67)

Although the preceding quote may be talking about the business imperative to "get with the Web 2.0 program," similar potentials for inclusion and shifting expectations around what constitutes expert knowledge also impact upon the conduct of research. Rather than academics seeing ourselves as experts sitting above society, increasing focus in community-based research projects should be placed on facilitating aggregation of the expert knowledge already present within communities. This process is already underway with the emergence of participatory GIS (PGIS) (see Elwood 2006a, 2006b; Elwood and Leitner 1998; Rambaldi and Callosa-Tarr 2000), and is starting to emerge in social sciences research where the knowledge of participants and their communities is respected in its own right, and thus research "subjects" enact agency, as active coresearchers (Tacchi, Hearn and Ninan 2004; Tacchi 2007). Such research is more commensurate with the kind of communal, open-source knowledge production processes enabled by digital technologies. This is content as process, not just endpoint.

CONCLUSION

"Space" and "place" emerged in the 1990s as key concerns for cultural researchers employing humanities and social sciences methodologies in their work, building on over a century's worth of work in geography on the spatiality of culture (e.g., Amin and Thrift 2002; Barcan and Buchanan 1999; Couldry and McCarthy 2004; Harrison, Pile and Thrift 2004). Despite this, it is only recently that GIS methods have come to be employed outside of geography and planning. Initial links to social science have been forged mostly by cultural geographers, and even then, frequently to those operating at the "hard" end of the social sciences: political scientists, criminologists, urban planners, and demographers. As the technology—both proprietary and open-source software—becomes more freely available, the benefits of GIS for analysis in the humanities and social sciences are its new ways to generate qualitative data that can be compiled systematically, its contributions to data visualization, and its ability to reveal otherwise subaltern patterns and voices. Alongside growing ideas of knowledge production enabled by networked digital technologies and a willingness on the part of some humanities and social sciences researchers to embrace visual methodologies in qualitative research (Banks 2007; Matei, Ball-Rokeach, and Qiu 2001), GIS mental mapping as part of an ethnographic project offers one exciting way forward. It offers a method that overcomes the inherent problems of relying on quantitative data alone to tell complex stories of connection, belonging, and place, not to mention of both formal and informal economic networks. As such, it also operates as a triangulation tool within complex, multimethod interdisciplinary studies. In the context of creative industries mapping, such an approach is an important step forward: one that can do justice to the complexity of the communities it studies and that allows researchers to enter into dialogue with these communities. Creative places call for creative research techniques, if we as scholars are to do justice to their worlds.

NOTE

1. For a full description of how the maps were administered during the interviews, refer to Brennan-Horley and Gibson 2009.

REFERENCES

Allen, R. C. 2008. *Going to the show: Representing the spatiality of film history.* Paper presented at the Centre for Critical & Cultural Studies, University of Queensland, Brisbane, Australia, March 11.

Amin, A., and N. Thrift. 2002. *Cities: Reimagining the urban.* Cambridge: Polity.

Australian Bureau of Statistics. 2004. *Experimental estimates and projections, Aboriginal and Torres Strait Islander Australians, 1991 to 2009.* Catalogue 3238.0. Canberra: Author.

Australian Bureau of Statistics. 2006. *Basic community profile—Darwin.* Canberra: Author. http://www.abs.gov.au/census (accessed April 16, 2007).

Banks, M. 2007. *Using visual data in qualitative research.* Thousand Oaks, CA: Sage.

Barcan, R., and I. Buchanan, eds. 1999. *Imagining Australian space: Cultural studies and spatial inquiry.* Perth: University of Western Australia Press.

Bauman, T. 2006. *Aboriginal Darwin: A guide to important places in the past and in the present.* Canberra: Aboriginal Studies Press.

Brennan-Horley, C. 2007. Work and play: The vagaries surrounding contemporary cultural production in Sydney's dance music culture. *Media International Australia* 123:123–27.

Brennan-Horley, C., and C. Gibson. 2009. Where is creativity in the city? Integrating qualitative and GIS methods. *Environment and Planning A* 41:2295–2614.

Bryman, A. 2008. *Social research methods.* New York: Oxford Unviersity Press.

Couldry, N., and A. McCarthy. 2004. *Mediaspace: Place, scale and culture in a media age.* New York: Routledge.

Cunningham, S., G. Hearn, S. Cox, A. Ninan, and M. Keane. 2003. *Brisbane's creative industries 2003.* Technical report. http://eprints.qut.edu.au/2409 (accessed April 30, 2006).

Drake, G. 2003. "This place gives me space": Place and creativity in the creative industries *Geoforum* 34:511–524.

Elwood, S. 2006a. Beyond cooptation or resistance: Urban spatial politics, community organizations, and GIS-based spatial narratives. *Annals of the Association of American Geographers* 96:323–41.

———. 2006b. Negotiating knowledge production: The everyday inclusions, exclusions, and contradictions of participatory GIS research. *Professional Geographer* 58:197–208.

Elwood, S., and H. Leitner. 1998. GIS and community based planning: Exploring the diversity of neighbourhood perspectives and needs. *Cartography and Geographic Information Science* 25:77–88.

Gibson, C. 2003. Cultures at work: Why "culture" matters in research on the "cultural" industries. *Social and Cultural Geography* 4:201–17.

Gibson, C., P. Murphy, and R. Freestone. 2002. Employment and sociospatial relations in Australia's cultural economy: *Australian Geographer* 33:173–79.

Gibson, C., and C. Brennan-Horley. 2006. Goodbye pram city: Beyond inner/outer zone binaries in creative city research. *Urban Policy and Research* 24:455–71.

———. 2007. *Creative tropical city: Statistical data analysis.* Report for Darwin City Council, NRETA, Arts and Museums Division and Tourism NT. Wollongong: GeoQuest Research Centre, School of Earth and Environmental Sciences, University of Wollongong.

Harley, J. B. 1989. Deconstructing the map. *Cartographica* 26:1–20.

Harrison, S., S. Pile, and N. Thrift, eds. 2004. *Patterned ground: Entanglements of nature and culture.* London: Reaktion.

Hearn, G., A. Ninan, I. Rogers, S. Cunningham, and S. Luckman. 2004. From the margins to the mainstream: Creating value in Queensland's music industry. *Media International Australia* 112:101–14.

Huynh, N. T., and S. T. Doherty. 2007. Digital sketch-map drawing as an instrument to collect data about spatial cognition. *Cartographica* 42:285–96.

Jull, P. 1991. *The politics of northern frontiers in Australia, Canada, and other "first world" countries.* Darwin: North Australia Research Unit, Australian National University.

Klenotic, J. 2008. MappingMovies.com. http://www.mappingmovies .com (accessed March 25, 2008).

LeGates, R. 2005. *Think globally, act regionally: GIS and data visualization for social science and public policy research.* Redlands, CA: ESRI.

Luckman, S., C. Gibson, and T. Lea. 2009. Mosquitoes in the mix: How transferable is creative city thinking? *Singapore Journal of Tropical Geography* 30:70–85.

Luckman, S. 2008. "Unalienated labour" and creative industries: Situating micro-entrepreneurial dance music subcultures in the new economy. In *Sonic synergies: Music, identity, technology and community,* ed. G. Bloustien, M. Peters, and S. Luckman, 185–94. Burlington, VT: Ashgate.

Lynch, K. 1960. *The image of the city.* Cambridge, MA: MIT Press.

Markusen, A., G. H. Wassall, and D. DeNetale. 2008. Defining the creative economy: Industry and occupational approaches. *Economic Development Quarterly* 22:24–45.

Matei, S., S. Ball-Rokeach, and J. Qiu. 2001. Fear and misperception of Los Angeles urban space: A spatial-statistical study of communication-shaped mental maps. *Communication Research* 28:429–63.

McRobbie, A. 2002. Clubs to companies: Notes on the decline of political culture in speeded up creative worlds. *Cultural Studies* 16:516–31.

Montgomery, K. C. 2000. Children's media culture in the new millennium: Mapping the digital landscape. *Children and Computer Technology* 10:145–67.

National Collaborative Research Infrastructure Strategy. 2008. *Review of the NCRIS Roadmap—Discussion paper.* Canberra: NCRIS.

Neff, G. 2005. The changing place of cultural production: The location of social networks in a digital media industry. *Annals of the American Academy of Political and Social Sciences* 597:134–52.

Neff, G., E. Wissinger, et al. 2005. Entrepreneurial labor among cultural producers: "Cool" jobs in "hot" industries. *Social Semiotics* 15:307–34.

Parker, R. N., and E. K. Asencio. 2008. *GIS and spatial analysis for the social sciences: Coding, mapping, and modelling.* New York: Routledge.

Peake, S., and T. Moore. 2004. Analysis of distortions in a mental map using GPS and GIS. Presented at SIRC 2004: A Spatio-temporal Workshop, proceedings of the 16th Annual Colloquium of the Spatial Information Research Centre, University of Otago, Dunedin, New Zealand, November 29–30. http://business.otago.ac.nz/ SIRC/conferences/index2004.html (accessed August 10, 2007).

Pratt, A. C. 1997. The cultural industries production system: A case study of employment change in Britain, 1984–91. *Environment and Planning A* 29:1953–1974.

———. 2004. Creative clusters: Towards the governance of the creative industries production system. *Media International Australia* 112:50–66.

Primorac, J. 2004. Mapping the position of cultural industries in Southeastern Europe. In *Cultural transitions in southeastern Europe*, ed. N. Svob-Dokic, 59–78. Zagreb, Croatia: Institute for International Relations.

Rambaldi, G., and J. Callosa-Tarr. 2000. *Manual on participatory 3-D modeling for natural resource management.* Quezon City, Philippines: National Integrated Protected Areas Programme (NIPAP).

Rose-Redwood, R. S. 2006. Governmentality, geography, and the geocoded world. *Progress in Human Geography* 30:469–86.

Scott, A. J. 2000. *The cultural economy of cities.* London: Sage.

———. 2004. The other Hollywood: The organizational and geographic bases of television-program production. *Media, Culture and Society* 26:183–205.

Shorthose, J. 2004. Accounting for independent creativity in the new cultural economy. *Media International Australia* 112: 150–61.

Steinberg, S.J., and S.L. Steinberg. 2006. *Geographic information systems for the social sciences: Investigating space and place.* Thousand Oaks, CA: Sage.

Tacchi, J. A. 2007. Ethnographic (per)versions and creative engagement through locally created content. Presented at CMS Symbols— Symposia on Communication for Social Development, Hyderabad, India, November 1–3. http://eprints.qut.edu.au/10390 (accessed November 5, 2007).

Tacchi, J. A., G. N. Hearn, and A. Ninan. 2004. Ethnographic action research: A method for implementing and evaluating new media technologies. In *Information and communication technology: Recasting development*, ed. Kiran Prasad, 253–74. New Dehli: B. R. Publishing.

Tapscott, D., and A. D. Williams. 2007. *Wikinomics: How mass collaboration changes everything.* New York: Portfolio.

Throsby, D. 2001. Defining the artistic workforce: The Australian experience. *Poetics* 28:255–71.

Tuan, Y. 1975. Images and mental maps. *Annals of the Association of American Geographers* 65:205–13.

Toohey, P. 2008. Darwin forced to reach for the sky. *The Weekend Australian*, June 14–15, 1–2.

Vajjhala, S. 2005. Integrating GIS and participatory mapping in community development planning. Presented at 25th Annual ESRI User Conference, San Diego, CA, July 25–29. http://proceedings. esri.com/library/userconf/proc05/abstracts/a1622.html (accessed February 14, 2007).

Watson, A. 2008. Global music city: Knowledge and geographical proximity in London's recorded music industry. *Area* 40: 12–23.

Upwave Cities, Creative Cities: The Case of London

John Montgomery

Urban Economist and Managing Director of Urban Cultures Ltd, London, United Kingdom

INTRODUCTION: THE ECONOMICS OF CITIES

The saddest book on cities I have read in a long time is *The Lost City of Stoke-on-Trent* by Matthew Rice (2010). Rice, who is married to the pottery owner Emma Bridgewater, charts the long decline of the potteries since the 1970s, when many brands closed local potteries to move overseas to Indonesia. There are now only a dozen or so potteries left in Stoke and many jobs that once were there have simply vanished. Yet at one time, Stoke was a place of great wealth creation, innovation and industriousness. The lesson is that once a local economy loses its dynamism, the place itself stagnates and may even die. Stoke is to the UK what Detroit is to the USA. Rice also shows that successive attempts at urban renewal have largely failed to make any impact in reversing Stoke's declining fortunes. Economic stagnation and decline occurs in real places, leaving multiple economic, social and cultural problems in its wake. Over a period of years, local communities and residents gradually grow poorer, as wealth leaks away to other places.

There are other examples: Liverpool and the decline of the docks, Huddersfield's shrinking wool industry, shipbuilding on the Clyde, rope-making in Newcastle, a large swathe of the car industry in the West Midlands, the cotton mills of Manchester and Oldham. All of these examples mark the decline of old industries, disinvestments, relocation overseas and – in the case of the docks – a switch to new facilities nearby as a response to union militancy. Cities can grow; they also can die.

And yet, there are also examples of other city-regions where these very industries are still dynamic and create wealth: Milan for textiles and fashion; Damietta for furniture; Copenhagen for electronics, pottery and jewellery; Prague for electronics and pharmaceuticals. Other cities have grown specialisms in new industries: biomedicine in Zhangjiang; Silicon Valley and Seattle for software; Hong Kong's film industry; and more. A recent analysis of seven European cities – Amsterdam, Munich, Barcelona, Leipzig, Birmingham, Manchester and Helsinki – concludes that cities grow in waves, building layers of wealth creation upon previous waves (Bontje *et al.* 2011). Each has in recent times benefited from new wealth creation with the emergence of new specialisms. That is all except Leipzig, a shrinking city that young people leave to better themselves in the cities of West Germany. Communism had destroyed the city's tradition of trade, printing and education. Leipzig's economy had virtually died under Communism (Gerkens 2005).

Amsterdam's great wealth had been built from the early seventeenth century and a second 'golden age' in the late nineteenth century: the city is now benefiting from a multi-layered diverse economy and society. Helsinki has been transformed by Nokia, the wireless communications giant, as well as small creative businesses housed in the Cable Factory and other managed workspaces. Munich also has a varied urban economy, with high-tech industry being a strong specialism. In the UK, Manchester is praised for its pioneering work by musicians, and others in the creative economy. Most of these cities were on the up from the late 1990s until 2008.

And so, just as economic decline affects real towns and cities, economic development and wealth creation also occurs in real geographic spaces. This reflects the very nature of economic growth. The truth is that economies do not grow gently on a steady trend; rather, they grow in leaps and bounds. Every 50 years or so capitalist economic development produces a wave of prosperity, as new processes, goods, markets and services open up to profitable investment. This becomes possible when new technologies are applied commercially, very often a few years after the original breakthrough has been made. One such wave is occurring now, particularly in digital media, mobile communications, applications and the internet (Montgomery 2007). Other leading growth sectors include bio-technology, marine biology, environmental technology and nano-technology, and the creative industries. Older industries – such as the potteries or the car industry – can and should restructure, specialise or otherwise adapt to changing patterns of demand and competition.

When cities catch the upwave of innovation, trade and prosperity, they generate wealth, employment and opportunities. Jane Jacobs (1961) developed a model for city economies, in which she demonstrated that cities achieve growth primarily through a process of exporting goods and services in order to earn surpluses with which to purchase imports. As the economy develops, more and more exports generate greater surpluses, during which time the division of labour among local producers becomes more complex. Various supply chains develop, so that networks of businesses provide inputs to the final export product. As people become wealthier, they can afford new goods and services, consumed locally, for which there had previously been no demand. They can also begin to make things themselves, and so replace imports with new locally based work. This results in economic specialisms, but also a build-up over time of networks of inter-trading businesses and 'clusters' of dynamic firms and whole industries.

The next stage is very important, for over time a strong and/or growing city economy will develop the skills and the capacity to replace or substitute imports and make these products locally. By this means, a city economy can then import other goods with its export surpluses and meantime the newly replaced imports may become another successful export product. Of course, during this time the city economy continues to export its existing goods and services, while the local division of labour becomes more diversified and complex. The wealth that is created is then re-invested in productive capacity, raw materials or stock; and a good proportion of this will also be spent by local citizens on consumer goods and services. The growth of local consumer spending in periods of rapid economic growth will closely be followed by, and in turn will create, increasing demand for new products – chocolates and coffee, fashions and jewels.

Jacobs goes on to demonstrate how these processes interlock and go on to produce multiplier effects for both exports (more local producers and exporters) and import-replacement (to meet local demand initially but later as new exports). Cities where these processes are self-generating and dynamic can enjoy periods of what Jacobs terms 'explosive growth'.

LONG WAVES OF CAPITALIST DEVELOPMENT

For the past 500 years, but most rapidly since the 1780s, western economies have developed in periodic spurts of wealth creation, artistic development, and physically as we build new cities. This process is as old as capitalism itself. Long periods of economic growth are followed by periods of stagnation and low growth, as night follows day. This occurs because the goods and services associated with particular phases of growth become exhausted, in the sense that there are no longer profitable markets to be sold into. The outcome is that particular goods and services – cotton in the eighteenth century, steel and transport in the nineteenth century, automobiles and electric goods in the early twentieth century, and consumer electronics in the late twentieth century – reach the limits to market penetration. Because of this the level of sales fall, prices are cut, the rate of profit falls, share prices fall, businesses fail, jobs are lost, and a cycle of decline sets in as aggregate levels of demand fall again. Economies operate along business cycles of 9, 25 and 60 years.

The most significant work on long wave business cycles was carried out by Joseph Schumpeter, at Harvard University in the 1930s (Schumpeter 1982 [1939]). Drawing upon the work of the pioneering Russian economist Nikolai Kondratiev, Schumpeter confirmed three 'Kondratiev cycles', each associated with the rise of new industries: from 1789 (the early industrial revolution); from the 1840s (the age of steam, steel, rail and ships); and from 1896 (electricity, motor cars and 'Fordism'). Schumpeter stressed that the important point is not the invention of a technology, but rather its applications to commerce, trade and production processes. He argued that economic growth is triggered by the emergence of new groups of entrepreneurs, prepared to take on new forms of work. He saw this as a sort of 'generational struggle' between old money and new wealth creators, or 'new men'. Schumpeter saw that economies, of necessity, grow in spurts because entrepreneurial behaviour lies at the heart of all economic activity. This, in turn, generates new applications for and combinations of technologies, using human skills to generate new industries and products. For Schumpeter, this is a process of 'creative destruction' where the new evolves from, replaces and supplants the old.

Writing in 1939, Schumpeter would have known that the next upswing – lasting roughly until 1974 – was about to happen. As it turned out, the technologies this time around were electronics, commercial aviation, consumer household goods, radio and sound recording. These technologies would become exhausted, to varying degrees, by the 1970s and 1980s, at which point new technologies were being invented and/or applied. Thus the period 1975–2000 roughly equates with a Kondratiev depression, during which time the general rate of profit fell, older industries declined, unemployment rose and new technologies – in this case computers, digital media, pharmaceuticals–were invented and began to be applied commercially. If this is true, then it seems likely that the capitalist economy in the year 2000 entered an upswing that will last for twenty-five to thirty years, that is until 2030 or thereabouts.

A recent important contribution by Carlota Perez (2002) argues, following Kondratiev and Schumpeter, that economic growth takes place in successive surges of about half a century, and these are driven by a technological revolution. She posits that there are six 'predictable' stages to these surges: technological revolution, financial bubble, collapse, golden age of prosperity and political unrest. According to Perez, each technological revolution leads to an investment frenzy as would-be entrepreneurs seek to get ahead of the game, and as they convince the financial sector to lend them money. This might seem irrational to some, but in truth each and every entrepreneur is acting very rationally indeed. She gives the example of the 'dot.com' boom and bust of the late 1990s, but the railways in the mid nineteenth century followed a similar path. The ensuing over-development and over-valuing of new investments – Perez characterises this as a 'divorce between paper and real value' – is followed by a crash, but this crash is short-lived as the economy generally is on a growth trajectory. The bubble and collapse is thus followed by a golden age of wealth creation as markets develop, capacity is increased and demand is increased. The final stage, political unrest, sets in as real inflation re-appears, productivity stagnates, full employment is reached and trades unions make unreasonable demands for wage increases. Precisely this happened in the late 1960s and early 1970s. As the economy moves into a long downswing, certain old industries begin to decline and unemployment increases. If Perez is correct, then the world economy is entering a phase of 'Synergy' or a long period of 'coherent economic growth based on increasing production and employment and a "re-coupling" of real and paper wealth'. This, then, is a golden age.

THE LONDON ECONOMY: A HISTORICAL PERSPECTIVE

London's economy re-invents itself continuously and has prospered in great waves for at least 500 years. London has in the past been a centre for electrical engineering, food processing and the car industry; today it is the world's largest centre for financial services and also has a large media and creative industries economy. In the wider London city region, industries such as pharmaceuticals, bio-technology, aerospace and defence-electronics prosper in places such as Bristol, Crawley and Stevenage.

Peter Hall has written convincingly of London's great golden age during Elizabethan times, that is from about 1570 until 1620, noting that 'something quite extraordinary was happening … London became a very special kind city' (Hall 1998: 114). Following a period of economic stagnation towards the end of the sixteenth century, London's economy and its population grew rapidly. Overseas trade increased rapidly, especially to the Americas and India, but also to the Baltic and Mediterranean. Domestic markets were also growing (within London but also with other parts of England), but it was the great voyages of exploration by the likes of

Drake and Raleigh that opened up new markets, new trade routes and brought new goods to England. A new class of entrepreneurs emerged, and the Dissolution of the Monasteries meant that land was available for development. Purchases of luxury goods from overseas – fashions, silks, coffee, tea – grew rapidly. New areas were opened up to housing for the new rich, notably in Southwark to the south and along and around the Strand in what is now London's West End. All manner of industries grew: leather manufacturing, glass, brewing, bricks and tiles, clocks, dye works, sugar refining and, of course, ship-building.

It against this backdrop that London's theatre industry grew, with people such as Ben Johnson, Inigo Jones, John Donne, Thomas Campion, Richard Carew, Raleigh, Marlowe and Shakespeare, and James Burage who built the first Globe Theatre at Bankside in 1598. The theatre flourished despite various attempts by the authorities to close venues and ban plays. It became a thriving industry, one that required large amounts of capital to be put at risk. It employed actors, set constructors, printers and costume-makers in previously unseen numbers. Its income came largely from admission prices, paid by audiences who had developed a taste for theatre. John Maynard Keynes would later argue that 'We were just in a position to afford Shakespeare just at the moment when he presented himself' and that 'by far the larger proportion of the world's great artists and writers have flourished in an atmosphere of buoyancy, exhilaration and the freedom of economic cares' (quoted in Knights 1937: 54). If London's dramatic growth of the early seventeenth century was indeed an upwave, that is a period of wealth creation lasting some 30 years, then one of its beneficiaries was the theatre and English culture. What was special about London was that it was at the forefront of economic advance, acting as a magnet for talent and entrepreneurs.

The good times would not return until after the Restoration, during and after which Britain (as it then had become) flourished as a result of increased trade with the Americas and with India. By the late eighteenth century London was a great trading port, an entrepôt, depot city and manufacturing centre rolled into one. All manner of port-related enterprises flourished, as did street markets specialising in meat, fish, fruit and vegetables, and livestock. Large amounts of food were imported from rural areas and from overseas, notably Denmark. There were all manner of specialised industries, for example in optics, watches and diamonds, the latter clustered around Hatton Gardens.

By 1800, Lombard Street had become the financial centre of world trade, 'paper' and 'names' oiled the wheels of commerce and the marine insurance industry, and the stock exchange had become established at Capel Court (White 2008). As the industrial revolution boomed across the country, the trade in securities and stocks grew rapidly. Despite financial panics such as the 'Spanish Panic of May 1835', or the collapse of Mexican and South American shares around the same time, the financial sector in London was on a long-term growth trajectory. A major reason for this growth was the emergence of the railway age, leading up to the 'railway mania' of 1854. By this time, London's economy was in the upswing phase of the second Kondratiev upwave. The volume of trade in the sector doubled on average every 5 years from the 1830s to the late 1860s. By 1866, some 170,000 people worked in 'the City'. This continued more or less unabated until the 1880s, despite events such as the Great Panic of 1857 when railway stocks collapsed in the USA and the panic spread to Europe and Britain (Montgomery 2011: 38–41).

It also during the second part of the nineteenth century that another of London's great industries matured: publishing. The industry began to flourish in the late eighteenth century, from its base around Paternoster Row, into Fleet Street (newspapers) and the West End more generally. By 1818 there were eighteen publishers located in 'the Row' and a further twenty-three in Fleet Street and the West End. As Jerry White (2008: 235) puts it, 'there was a vast expansion of newspapers, reviews, journals, magazines and cheap novels' in London during the eighteenth century. Many writers were attracted to publish, notably educated middle class women such as Annie Besant, men in the civil service, professionals such as Somerset Maugham and Arthur Conan Doyle, and lawyers such as Wilkie, Collins and Thackeray. Not to mention Dickens. By the late nineteenth century London was home to the Bloomsbury set, and writers such as Henry James, Barrie, Oscar Wilde, Shaw and Aubrey Beardsley. Writers, many of them from the provinces or Ireland, were attracted to London by the prospect of earning money from writing, and by the real live experiment in the clash between tradition and modernity, the old and the new. And, because of advances in the technologies of printing and distribution (rail, steam ships), there was a strong demand for writing.

With this hunger for the written word, there followed a growing demand for theatre and entertainment. By the late 1880s, the West End was firmly established as London's theatre-land (Weightman 1992). Shaftsbury Avenue was opened in 1888, and quickly became the city's main theatrical street, with a boom in theatre building that lasted until 1910. New customers were drawn from the suburbs, easily accessible by train and underground. In this way, during the second half of the nineteenth century, and in response to demand – itself created by two waves of wealth creation – London developed two major 'creative' industries: publishing and commercial theatre. Both of these were market economies.

In the early part of the twentieth century, the music halls began to be replaced by cinemas. The first West End cinema, the New Egyptian Hall in Piccadilly, opened in 1907. By 1912 there were 500 cinemas in London. The development of sound recording led to 'the talkies', and from the late 1920s there was a new boom in cinema development. In 1928, the old Empire in Leicester Square was converted to a cinema with seating for 3,000 people. A new industry had developed: film production.

By the early part of the twentieth century, London was again caught up in a wave of wealth creation. This time the new lead industries were food processing, armaments, electrical engineering and the film industry. Centuries of growth, innovation and investment had produced an astonishingly diverse economy. London brewed its own beer, had specialised markets such as Spitalfields and Smithfield scattered around the city, had large numbers of wharves and related activity, the manufacture of printing presses and machinery, ink and paper, and much more. Clothing was manufactured in Hackney and Whitechapel, employing some 250,000 people in 1901, although boots and shoes were imported from Northampton and Leicester. London also had a large furniture industry based in Shoreditch and Bethnal Green. There were also strong exporting industries in scientific instruments, electrical instruments, light bulbs and musical instruments. At this time, London was known as the 'Mart and Workshop of the World' (White 2008). There was also heavy industry such as van and cart works, chemical works, bottle-makers, canvas factories, shipbuilding and foundries, located in the East End near the docks.

West London was from the 1920s, and again from the 1950s, the UK's fastest growing industrial district. London had grown from a trading port and an entrepôt and market, to a centre for financial services and publishing, to a manufacturing city and then a great industrial district. The new industries of the twentieth century sought space to locate and expand: activities such as electrical engineering (Marconi), lighting (Lee), pharmaceuticals (Glaxo), Ilford Photographics, household appliances (Hoover), food processing (Heinz, McVities) and brewing clustered in places such as Park Royal, Acton, Willesden, Alperton and Wembley. Park Royal grew from eighteen factories in 1918 to 250 by 1939, and factories continued to open even during the Great Depression. So many jobs were created that, at the height of the Depression, West London had a severe labour shortage. This produced a built form characterised by large factory buildings (many of them art deco), set in large grounds with easy road access and ample room for storage. The film industry developed at large studio complexes in places such as Wembley, Ealing, Pinewood, Teddington and Shepperton. This situation emerged because the studios needed grounds in which to build studios and sets. Suppliers of electrical equipment and film were also easy to hand. Meanwhile, the BBC developed its own large studio complex at White City in West London.

In this way, what Schumpeter referred to as the Third Kondratiev Wave led to a radical change of work in London, with new products but also new forms of factory-based and mechanised production. Advances in transport technologies (the automobile) opened more parts of London up to industrial development, while the railways and underground enabled growing numbers of bankers, city workers, office workers and civil servants to commute from Surrey, Middlesex and 'Metro-land'. The 'creative industries' of the day were publishing and cinema, plus the West End Theatre, and these too experienced a boom.

But by the late 1950s, cinema and film production were coming under pressure for audiences from a newly commercialised technology, television, initially in the form of the BBC. In the 1950s too, traditional industries such as clothing and furniture were losing jobs. On the positive side, the growth of commercial aviation led to a great expansion of airport and related activities at Heathrow, where passenger traffic grew from 282,000 in 1947 to 5.3 million in 1960. Air freight also expanded rapidly.

Meanwhile, back in central London, manufacturing was being forced out by office buildings and the growth of the service sector, and by the development of new retail centres. This was underpinned by a short-sighted policy by London County Council to encourage manufacturing and craft industries to move out of London. In addition, port owners developed new facilities at non-unionised locations such as Harwich and Felixstowe, while the development of motorways to the north and west encouraged many of the great industrial firms of West London to move further away from the capital. The growth industries in London at the time were banking and finance, commercial aviation, tourism and retailing, and office work in general, but also television, radio, sound recording and distribution, and advertising. The Fourth Kondratiev Wave witnessed a growth in some industries in London, but also a decline in dock-related industries and manufacturing. By the onset of the downswing, in the late 1970s, London was on the way to losing 36 per cent of its manufacturing jobs. The worst year for job losses was 1994, following the stock market collapse of 1987.

By the early twenty-first century, it was estimated that the creative industries employed more than 525,000 people in Greater London, and that the creative industries in London had a combined turnover of £21 billion (GLA 2002). The growth in output for the period 1995–2000 was 8.5 per cent per year so that, by the year 2000, the creative industries were the third largest economic sector in London. Collectively, the total output of the creative industries was greater than any other industrial sector except business services, with creative industries output exceeding financial services, manufacturing, health and education, transport and retail. Advertising, architecture, art and antiques, software and computer games, and film and photography all showed particularly strong growth rates, above the average rate for all of London's creative industries. In 2002, the creative industries workforce in London was 468,700, including those directly employed in creative industries sub-sectors, plus self-employed workers. An additional 182,000 people had creative jobs in London, but also in sectors outside of the creative industries (e.g. designers for car manufacturers, musicians in the education sector.) The creative industries are the third largest employment sector in London, after business services, and health and education. Between 1995 and 2000, creative occupations were London's second biggest source of job growth, increasing by 111,000 and contributing roughly one in five new jobs.

Key London creative industries locations include Soho and the West End, but also places such as Camden (film and TV production), Fitzrovia (television production and fashion design), Clerkenwell (crafts) and North Kensington (film and video production and music recording). The chief design – particularly graphic design – cluster is alongside advertising and the other media industries in Soho and Fitzrovia and also in Covent Garden. Textile or fashion designers cluster in the rag trade area of Fitzrovia in and around Great Titchfield Street. There is also a cluster of more 'alternative' designers around the Kings Road. This was the heart of the punk street fashion explosion of the 1970s. Other small clusters (more to do with graphic and product design) can be identified around London Bridge (SE1), around Hammersmith Bridge (W6) and around the Portobello Road area (W10). Many locate in managed workspaces, for example the Barley Mow in Chiswick and any number of similar buildings in Covent Garden.

In fact, it was evident by the mid 1990s that the creative industries accounted for just over 6 per cent of London's total employment, some 215,000 jobs. For most of the creative sectors – especially advertising, video and the music industry – there had been a growth of employment during the 1980s, but falling employment levels from the late 1980s until about 1995. Changes in employment levels were partly due to the impact of economic recession of 1988–1994, but also to industry restructuring, and the ongoing process of mergers and takeovers (Urban Cultures 1994).

In this way, the growth of new industries during the early part of a Kondratiev upswing appears to have occurred in the late 1990s and early 2000s, particularly for the creative industries in central London, but also in pharmaceuticals, defence electronics and high-end R&D in the London city region, which these days encompasses places such as Aldermaston and Stevenage. If this is the case, then the creative industries are one of the lead industries of the current upwave. Because of this, London continues to attract the talented, the bright and the ambitious. However, there are already signs that only some creative industries are new and will continue to expand, while some, notably cinema, television and the music business, are struggling against greater competition, the internet and changing consumer choices. London's economy has changed in the past, usually, over time, for the better. It has caught a good share of new industries along each long wave of technological progress and wealth creation. It is doing so again.

Jane Jacobs argued that city and city region economies expand in bouts of phenomenal growth (Jacobs 1969). She also argued that economic development is a version of 'natural development' (Jacobs 2000), such that economies only succeed where there are economically creative and innovative people. Economic development is a non-linear system; economies make themselves up as they go along. Diverse economies expand in a rich environment, which is itself created by diverse use of local resources and imported materials. A 'gene pool' of work builds up over time of knowledge, skills, techniques and methods. This leads to a type of 'creative self-organization' and a continual and continuing process of adaptation and improvisation. Development produces diversity, but also depends on co-development of related elements, and because of this development is open-ended not pre-planned. Growth depends on trade and innovation. Successful dynamic economies alter and change by a process of 'aesthetic drift', moving in response to myriad influences, signals and technologies.

Dynamic economies are based in growing cities and regions: economic growth is a product of trade between city-regions. To maintain growth and levels of wealth creation, economic systems need regular infusions of energy. In cities and city-regions, this comes from new work and imports from which further new work is in turn created (as imports are replaced to be produced locally). A portion of received energy must be devoted to capturing new energy, new imports. What Jacobs (2000) referred to as 'self-fuelling' is a process whereby cities generate their own new work, and thus future rounds of creation and growth. This is what London does so well. The important measurement of economic growth is not therefore aggregate levels of GDP across a country, but rather the economic health of key growth cities and their regions. At the local city-region level, the key indicator of economic health is the ratio of total economic value to the value of imports.

As in previous waves of prosperity – Manchester and cotton, Glasgow and Chicago and rail, Detroit and automobiles, Seattle and commercial aviation – certain cities enjoy a large share of the new wealth. Milan is a great centre of fashion and design, Helsinki has developed its mobile phone industry almost from scratch. Bangalore is a centre for IT and software development, Manchester has reinvented itself as a creative city. China, meanwhile, has developed several creative industry clusters and networks in city regions such as Foshan, Songzhuang, Dafen and Suzhou (Keane 2012). The cities that will prosper most in the coming period are already placed to do so.

And perhaps this is the main lesson: some cities and city-regions will do better than others in the new period of prosperity, because they are already successful, have traditions of knowledge and skills, offer export specialisms and are innovative enough to grow new work from old. Other cities might well lose out completely if they have lost economic diversity, know-how, and the capacity to innovate and invest. All the social enterprises in the world cannot make up this deficit, because the harsh fact is that economic development differs from community development, with the latter being designed to deal with particular socio-economic problems: urban blight, crime, housing renewal. These are not of themselves economic development in any sustainable sense. Only real economic development can underpin the wealth and prosperity of city regions. Sadly, on this reading, Stoke is a city region with insufficient economic diversity. If true, this means that the only way for cities in the North of England to prosper is for them to develop economically though innovation, investment and entrepreneurialism.

The cities that will do best in the coming years will have certain characteristics. Leading cities will have diverse economies and a high level of high value production in goods they export to other city regions (Fruth 2011). Because much of the new wave is based on industries with a high R&D component, requiring highly skilled people, most of them will be 'creative *milieux*' (Törnqvist 1983). These share four key features: information (which must be exchanged and inter-traded); knowledge (bodies of work and data-bases); competence in certain activities; and creativity, which combines the other three features to create new products, ideas and processes. In this way, creative places have a set of characteristics that, in most cases, take a long time to evolve and develop, often building upon previous waves of wealth creation. It takes time to build up libraries, archives, databases and traditional skills. Such places come to have a recognised set of specialisms, and these in turn act as a magnet to attract further generations of artists, designer, technicians and entrepreneurs.

The leading cities of the coming period will thus:

• have some share of the lead industries of the Fifth Kondratiev Wave

- be globally connected as centres of trade

- be adept at adaptation and innovation

- have diverse and multi-faceted economies

- have industries which are organised in complex groupings of clusters, networks and sub-sectors

- will have strong creative economies and sophisticated arts scenes

- will have sophisticated local patterns of consumption, and

- their urban form will provide a good 'fit' to the city's economy and cultural life.

They will be comprised of clusters of industries, networks of firms and individuals, social relations and cultural life. They will be self-generating urban and city-regional economies. They will be like London.

REFERENCES

Bontje, M., S. Musterd and P. Pelzer. 2011. *Inventive city regions: Path dependence and creative knowledge strategies*. Aldershot: Ashgate.

Fruth, W. H. 2011. *The flow of money and its impact on local economies*, National Association of Industrial and Office Properties, January 2011. www.naiop.org.

Gerkens, K. 2005. *Strategies and tools for shrinking cities – the example of Leipzig*. Paper presented to the International Cities, Town Centres and Communities Conference, June 2005, Capricorn Coast, Queensland.

Greater London Authority. 2002. *Creativity: London's core business*. London: GLA. http://www.london.gov.uk/mayor/economic_unit/docs/create_inds_rep02.pdf (accessed 30 January, 2012).

Hall, P. 1998. *Cities in civilization: Culture, innovation and urban order*. London: Phoenix Grant.

Jacobs, J. 1969. *The economy of cities*, London: Jonathan Cape.

Jacobs, J. 2000. *The nature of economies*, New York: Vintage.

Keane, M. 2012. *China's new creative clusters: Governance, human capital and investment*. London: Routledge.

Knights, L. C. 1937. *Drama and society in the age of Johnson*. London: Chatto & Windus.

Montgomery, J. 2007. *The new wealth of cities: City dynamics and the fifth wave*. Aldershot: Ashgate.

Montgomery, J., 2011. *Upwave: City dynamics and the coming capitalist revival*. Farnham: Ashgate.

Perez, C. 2002. *Technological revolutions and financial capital: The dynamics of bubbles and golden ages*. Cheltenham: Edward Elgar.

Rice, M. 2010. *The lost city of Stoke-on-Trent*. London: Frances Lincoln.

Schumpeter J. 1982 [first published 1939]. *Business cycles: A theoretical historical and statistical analysis of the capitalist process*. Philadelphia: Porcupine Press.

Törnqvist, G. 1983. Creativity and the renewal of regional life. In *Creativity and context: A seminar report*. Lund Studies in Geography, ed. A. Buttimer, 91–112. Lund: Gleerup.

Urban Cultures Ltd. 1994. *Prospects and planning requirements for London's creative industries*. London: London Planning Advisory Committee.

Weightman, G. 1992. *Bright lights, big city*. London: Collins and Brown.

White, J. 2008. *London in the 19th century*, London: Vintage Books.

Developing a Creative Cluster in a Postindustrial City: CIDS and Manchester

Justin O'Connor and Xin Gu

Creative Industries Faculty, Queensland University of Technology, Brisbane, Queensland, Australia

This article takes the establishment and demise of Manchester's Creative Industries Development Service (CIDS) as an exemplary case study for the ways in which creative industry policy has intersected with urban economic policy over the last decade. The authors argue that the creative industries required specific kinds of economic development agencies that would be able to act as "intermediaries" between the distinct languages of policymakers and "creatives." They discuss the tensions inherent in such an approach and how CIDS attempted to manage them and suggest that the main reason for the demise of the CIDS was the domination of the "economic" over the "cultural" logic, both of which are present within the creative industries policy discourse.

In the last twenty years the image of Manchester has gone from that of a declining industrial to a re-invented "postindustrial" city. Central to this new image have been both the subsidized arts—with major investment in infrastructure and events—and more commercial popular culture, most notably popular music (Haslam 1999). This narrative in which arts and popular culture become emblematic of a transformative "spirit of place" is now well established within "celebratory" texts on the city (Manchester City Council 2002, 2004; King 2006) and more critical and academic literature (Peck and Emmerick 1992; O'Connor and Wynne 1996; Taylor et al. 1996; Wynne and O'Connor 1998; O'Connor 1998; Brown et al. 2000; Ward 2000, 2003; Peck and Ward 2002; Haslam 2005; Hetherington 2007). Since 1998, when the UK Department of Culture,

Media and Sport (DCMS) launched a new definition and set of policy directions (DCMS 1998), this culture-led regeneration narrative has also used the term "creative industries" alongside "arts and culture."

Manchester has the largest creative industry sector in England outside of London, and its creative industries sector continues to grow as those of other cities stutter (NWDA 2008). Indeed, the BBC recently chose the city as the site for a large-scale relocation of production facilities from London to a new "MediaCity" in Salford Quays (http://www.mediacityuk.co.uk). This article looks at an earlier initiative, the Creative Industries Development Service (CIDS), a small not-for-profit agency set up in 2000 by Manchester City Council in partnership with a range of other local public agencies. Its goal was to promote the economic development of the creative industries by helping frame and providing a coordinated response to the needs and aspirations of the sector. Our account of CIDS highlights some of the tensions within local creative industry development policy in the last twenty years. Central to these tensions was a persistent ambiguity between "culture" and "economics" that still marks the wider academic debate (Hesmondhalgh 2007; O'Connor 2007; Banks and O'Connor 2009).

We suggest that the "creative industries," though often defined in the placeless language of the "knowledge economy," "creativity," and "innovation," are rooted in and held to exemplify complex local histories and cultures. We also suggest that the shift to "creative industries" has positioned the "economic" as the master signifier in these local policy discourses, in a way that frequently misunderstands the actual practices of the creative industries and cuts across their cultural identification with, and investment in, the city. We start by looking at the trajectory of cultural and creative industries policy discourse in a local economic context, and in particular the close connection to discourses of "culture-led urban regeneration." We then discuss this connection to place in terms of creative industries and policy in Manchester, before looking in more detail at CIDS. We focus in particular on CIDS attempts to encourage local creative networks, and how this brought out some of the wider tensions between culture and economics to which we refer. We end with some concluding remarks on local creative industries policy.

CULTURAL INDUSTRIES AND CULTURE-LED REGENERATION

Though the creative industries are often presented in terms of ubiquitous creativity and global communications (Howkins 2001; Hartley 2005), they are very much rooted in particular places. These complex local roots have been clearly identified by economic geographers working on clusters, embeddedness, and "cultural products industries" (e.g., Scott 2000, 2004), but this has been less recognized in terms of local policymaking. "Policy transfer" has been central to the proliferation of cultural consultancy firms since the 1980s, where the possibility of adapting successful ideas from elsewhere is the key to the innovative capacity of "creative cities" (Landry 2000). But creative industries policy discourses, even when ostensibly derived from a single source such as the DCMS strategy document, are very much local-path dependent (Kong Gibson, Khoo, and Semple 2006). Andy Pratt (2009) has usefully discussed the limits of policy transfer and the specificity of particular policy regimes at national-regional level, but this has rarely been applied to particular cities or regions.[1] If the "creative class" is supposedly highly mobile, looking to move to those cities that offer the best lifestyle infrastructure (Florida 2005), this is much less so with policymakers, whose "political capital" is very much linked particular places.[2] This is certainly so in the United Kingdom, where "creative industries" policy discourses arrived at least a decade after those of "culture-led regeneration" and "cultural industries."

The increased centrality of "arts and culture" to the "regeneration" of older industrial cities from the late 1970s onward is well-established. For critics it represents a "cultural fix" for global capital (Harvey 1989), organizing city centers around a process of gentrification and (cultural) consumption-based "landscaping" (Zukin 1982, 1991, 1995), resulting in a fantasy city (Hannigan 1998) or phantasmagorical cityscape. For others it had a more complex provenance in European urban cultural policy (Wynne 1992; Bianchini and Parkinson 1993; O'Connor 1998). Concern for "quality of place" expressed through cultural facilities and activities has certainly been a central concern of many—especially industrial—cities as part of an attempt to promote a new image to attract investment, mobile professionals, and tourists, as well as retaining existing residents. Although frequently dismissed as local "boosterism," many of these attempts were part

of a much wider process of economic and sociocultural change, which relatively more permeable "urban growth coalitions" (see later discussion) attempted to set within a shared narrative of a dynamic "spirit of place" capable of facing a new round of global competitive challenges. These urban narratives were not completely imposed by city marketing agencies but could tap into widespread (though often vaguely formulated) aspirations for the city on behalf of local citizens. These narratives, although primarily economic in traditional "boosterism," also had a strong cultural component—cultural not just in the sense of local identity but also in the more restricted sense of arts and popular culture. The popularity of "arts impact" studies may have declined since the 1980s (Myerscough 1988), but there is little doubt that even after a decade of "creative industries" it is "arts and culture" events and venues that remain the keystone of local authority "regeneration" narratives—even in a city where "creative industries" have grown significantly, such as Manchester. That is, despite their close alignment with narratives of economic transformation and "regeneration," they still carry a strong "cultural" charge. These local, cultural aspirational narratives, we shall argue, are particularly attractive for the "creative industries."

The increasing importance of "cultural industries" to UK local authorities in the 1980s needs to be situated within wider processes of sociocultural and political change, rather than dismissed as a "cultural fix" at the service of "capital," or indeed, of a local state incapable of influencing anything else (Peck and Ward 2002). Neither can their growing prominence within local cultural strategies, giving rise to a range of new consultancy firms (especially Commedia), be understood simply as a response to decreased funding for the arts (Hewison 1997); it relates to more complex notions of cultural democracy and local economic development that go back to the Greater London Council (GLC) of 1981–86 (Bianchini 1987; Garnham 1983/1990). The GLC vision, mostly embryonic, tried to deal with the fact that the vast majority of cultural consumption was delivered by the market, not by the subsidized sector; any democratic policy must confront this. It also drew on European urban policy developments, especially from Italy (Bianchini 1987), but also, as we shall see, new kinds of local economic strategy. It was both an economic and a cultural strategy, where countering the domination of (often large, multinational) distribution over (often local and small-scale) production meant having to rethink the "arts welfare" model of grants to individual artists. It moved to adopt an "industry model" approach—"industry" here referring to a designated economic sector, not mass production (O'Connor 2000)—with a concern for supply chains and complex combinations of creative and noncreative skills and businesses.

It is significant, however, that the first adoption of the notion outside of London was as an area-based regeneration strategy in Sheffield—the Cultural Industries Quarter (CIQ).[3] Although explicitly geared to creating the future jobs of a post–steel industry city, it was embedded in local strategy not through a citywide program but as a vehicle for regenerating a decaying city center site. To some extent it formed a part with developments in the Albert Dock, Liverpool (itself modeled on the Boston Waterfront development), Museum of Science and Industry in Castlefield, Manchester, and related heritage and niche retail projects. The area-based focus of CIQ had pragmatic policy reasons. There was little knowledge of how to promote the cultural industries at a local level, and such knowledge—or aspiration—as existed was closely linked with particular individuals who applied their energies to the limited policy base over which they had some control.[4] But there were wider political reasons, in that both local authority power and the wider rationale for local economic strategies were severely curtailed by the Thatcher government, elected in 1979.

Creative industries policies have been linked to "information society" discourses originating with Daniel Bell and Alvin Tofler in the 1970s (Granham 2005; Pratt 2009; Pratt and Jeffcut 2009). This certainly contributed to a discourse in which cities, liberated from the physical constraints of a manufacturing economy, seemed able to reinvent themselves through "creativity." But there were other policy narratives drawn from European social democracy that were very much at play, from the GLC initiative up to and beyond New Labour's redefinition. Indeed, much of early New Labour's economic thinking was concerned with a "third way" between state control and free market, and thinkers looked to the successful European regions such as the "Third Italy" and Baden-Wurttemberg. The emphasis here was on locally embedded networks of small- and medium-sized enterprises (SME) operating very differently from the vertical and hierarchical integration of (mostly national) mass manufacturing and the globally mobile multinationals with little commitment to place. These ideas seemed to provide an alternative to Thatcherism among the social-democratic left in the 1980s and 1990s (Piore and Sabel 1984; Hall and Jacques 1989; Zeitlin and Hirst 1989; Hutton 1995; Redhead 2004). However, the space in which such alternative local economic strategies could develop was limited by Thatcher's breaking up of the large metropolitan authorities (reducing Manchester City, for example, from a population of 2.5 million to just over 500,000) and by a series of legislative measures designed to restrict their powers of taxation, spending, intervention, and regulation. "De-industrialisation" was used to drive through a program of economic, social, and cultural reconstruction in a way unique in Europe. As a result, UK local authorities remain bereft of the industry intelligence and range of policy tools and powers deployed by their counterparts in other areas of Western Europe.

From the perspective of cultural and creative industries policies, what mainly survived were the notions of "cluster" and "creative milieu." The notion of "industry cluster" emerged into the mainstream of UK local economic policymaking in the late 1990s. This notion, frequently attributed to Michael Porter (1998, 2000) and much debated (cf. Martin and Sunley 2003), has a complex provenance in the work of social and economic geographers concerned to identify those unique, often intangible, qualities of place that gave rise to competitive advantage (Granovetter 1985; Piore and Sabel 1984; Powell 1990; Markusen 1996). Alfred Marshall's (1898) notion of "atmosphere," referred to by Porter, has been interpreted as the particular social and cultural dimensions within which local economic activity is embedded. If, in the United Kingdom, "cluster" merely came to refer to a local industry sector or subsector—"digital cluster," "music cluster," "bio-tech cluster," and so on—on the other hand it directed attention to the locational dimension of these industries and the "untraded interdependencies" that helped them thrive (Storper 1995). Economic geographers working in this field sought further to distinguish creative industry clusters from other business clusters, and emphasized the sociocultural dimensions of place as key factors of "competitive advantage." In so doing they have used terms such as innovative or creative milieu, creative field, or creative, critical, or soft infrastructure (Hall 2000; Pratt 2000, 2002, 2004; Scott 2000, 2001, 2004, 2006).

There were thus clear linkages between the notion of cultural and creative industry cluster and the more established models of cultural or creative quarters, districts, precincts, and the like. These latter had also developed within the culture-led urban regeneration agenda of the 1980s, and involved a range of different models unified around the concentration of some form of cultural production or consumption (Bell and Jayne 2004; Mommaas 2004; 2009; Roodhouse 2006; Montgomery 2007). They stressed the benefits of colocation for both production and consumption; the mix of public and private actors; diverse leisure, retail, and entertainment offers; and a wider concern with their contribution to, and benefit from, the image of the city within which they were located. Although members of the cultural and creative industry cluster might focus more directly on the commercial businesses within it, they clearly benefited from the "buzz" of consumption activities, and were similarly implicated in the wider "brand" of the city. Indeed, cultural and later creative industry strategies continue to be focused on building-based developments; while arts and museums get iconic modernist buildings, creative industries are still linked with industrial-era heritage. These latter developments are also promoted as iconic for the city's contemporary creative brand.[5]

CREATIVE INDUSTRIES AS SME POLICY: THE PROBLEM OF REPRESENTATION

UK cultural and creative industries policies remained very much linked to urban regeneration initiatives at least until the more recent government initiatives around the (re) launch of the creative industries in the DCMS's Creative Economy Programme in 2005.[6] Outside of regeneration policy, the main source for development policies was support programs for SMEs. Local economic development based on encouraging SMEs was derived from both the reinvented social economy ideas of the "Third Italy" and European social-democracy previously noted, and the more "neoliberal" or "New Right" ideas aimed at freeing entrepreneurship from bureaucratic obstacles. Thatcher's government in the United Kingdom was firmly committed to these as part of the "enterprise culture" and provided a whole range of legal and tax incentives for small businesses, along with an increase in training and business support. Many local governments seized on this agenda even if they had more leftish notions of local economic development. It was this general policy approach, especially the Enterprise Allowance Scheme, that provided a launch pad for many small-scale cultural activities in the late 1980s and early 1990s (McRobbie 1998; O'Connor and Wynne 1996).

Research reports from the time of the GLC on had indicated high levels of SMEs in the cultural industries sector and their concentration in metropolitan regions (Pratt 1997). As such, cultural and later creative industry strategies almost always had a strong focus on support for SMEs. But these reports also showed that standard forms of support—business advice and training, access to loans and startup grants, tax incentives, and the like—did not work for the cultural industries. Creative businesses were seen to have different ambitions, priorities, and ways of operating than 'mainstream" businesses (O'Connor and Wynne 1996; O'Connor 1998; O'Connor 2000; Banks Lovatt, O'Connor, and Raffo 2000; Raffo, Banks, Lovatt, and O'Connor 2000). Generally speaking, traditional business support agencies saw these cultural SMEs as "risky" businesses, and the cultural businesses themselves saw these mainstream agencies as having no real understanding of what they did. Cultural businesses operate in a speculative, risky and volatile environment, one not managed using formal market analysis or risk assessment but often by informal, "intuitive," aesthetic and ethical procedures. To many in mainstream business support, these sector-specific practices were failures because of lack of business training and could be remedied by generic "startup" programs. To some extent the specific problems faced by

these businesses were compounded by very basic generic errors—not understanding tax regulations or simple accounting, intellectual property and contract law issues, employment regulations, and so on. But even these problems had specific inflections in the cultural industries, and the solutions provided by mainstream support were often rejected by cultural businesses. The fundamental divide between cultural businesses and other businesses was often (though simplistically) encapsulated by the notion of "T-shirt and Suits" (Parrish 2005).

Despite this, from the mid-1990s there was a growing demand for some form of business advice and support among local cultural industries. A number of small specialized agencies started by giving different levels of business advice and training to sole traders and small cultural businesses. These were often subcontracted out by mainstream training and economic development agencies. Some local universities and colleges also began to provide bespoke training for prospective cultural entrepreneurs. Parallel to these, the Regional Arts Boards (the main channel of state subsidy to arts and cultural activities) also began to run business-oriented seminars and short training courses. After 1998 the DCMS also called for skills training and business support initiatives from its national "art form" agencies, and encouraged schools and further/higher education institutions to participate in creative industry focused training and skills programs. These were usually focused around promoting "creativity" and entrepreneurship (Redhead 2004; Bilton 2006; Schlesinger 2007; Banks and Hesmondhalgh 2009). However, at local level, business development and training for the creative sector developed on a much more ad hoc basis; this partly reflects the "motley" nature of creative industries voices and the question of policy leadership in an emergent field.

What kind of body was to "own" the creative industry agenda and how would one identify those who could be interpolated as subjects or agents for policy discourse? Large and more established industries (architecture, printing, broadcasting, film, classical music, theater, etc.) had their own industry, professional, and union bodies, often able to lobby governments directly. This was not so with the dispersed and diverse individuals and SMEs who made up the emergent creative industries "constituency." As far as this constituency did exist for national and local cultural policy agencies, it did so as part of the "art form" structure of the UK Arts Council and its regional offices. Specific art form bodies were set up to represent and channel funding for visual arts, theater, film, music, literature, craft, design, etc. The DCMS "mapping document" (1998) did add other subsectors (e.g., leisure software design) to the list of creative industries—but the problem was more than a simple extension of subsector agencies but one of the legitimacy and purpose of "representation." The "music officer" of the Regional Arts Board could not really speak for the local popular music scene, given the overwhelming focus on subsidies for "classical" music and its explicit disavowal of any involvement in commercial music activities. Computer games, on the other hand, represented a completely new field in which arts boards would have little or no expertise—nor would they want to acquire this.

Questions around commerciality and technical competences were part of a wider issue of who should lead on creative industries policy—existing cultural agencies (however suitably revamped) or local economic development agencies? A related question concerned whether creative industries should be seen as a series of subsectors to be led by specific agencies, or be constructed as a unified sector to be dealt with as a whole. Many of those in the creative industries did not, or refused to, see themselves as "creative industries"; indeed, one task of policymakers had been to create this sense of belonging and thus simplify and coordinate the policy delivery channels. They wanted the sector to "speak with one voice."

How were existing agencies to deal with this new object called the "creative industries?" As a first step, agencies had to identify appropriate individuals, organizations, and businesses with whom they could work and elaborate new strategic directions. This involved finding a representative "voice" from the sector and subsectors and identifying or creating an agency that could then work with this "voice" to deliver policy initiatives in a convincing way.

There were three problems here. First, existing policy agencies were not clear among themselves of exactly what the "creative industries" consisted; statistical and definitional questions were the relatively easy end of a complex conceptual issue, still not resolved (O'Connor 2007). Second, as it involved a new policy object, existing cultural organizations and economic development agencies had to work with "intermediaries" who were relatively unknown and untested. Popular music, having no local industry representatives, was notorious in this context—where musicians (the successful and the less so) were often turned to for complex policy advice (with varying degrees of success). Third, because of the strong "cultural" dimension to the creative industries, many businesses within this sector considered themselves to be something other than "industries" subject to economic development policy; those who had fewer reservations about this aspect often saw their form of creativity to be operating in direct opposition to "policy," "bureaucracy," and "institutionalization." The image of "Cool Britannia," mixing politicians and rock stars, was persistently invoked precisely as an image of "uncool."

However, there was a strong feeling among many creative businesses, certainly in Manchester but reflected in other UK cities, that they were being "misunderstood" or ignored by the local authority. Creative businesses were usually not reflected in the strategic or promotional literature of economic development agencies until the late 1990s. Many in the creative industries felt a lack of "representation," but this was less in terms of business support (about which many were skeptical, or oblivious) than in terms of their own contribution to, and stake in, the transformation of the city.

CREATIVE INDUSTRIES AND URBAN REGENERATION IN MANCHESTER

Part of the story of cultural and creative industries in UK cities (and elsewhere) is their impact on the urban landscape. As we have noted, it was the impact on the physical regeneration of the city that was the most prevalent strand of policymaking in the 1980s and 1990s. This was more than occupying and renovating old buildings; they changed the symbolic value of these buildings and the surrounding areas (Zukin 1981, 1992, 1995; O'Connor 2004). They made areas "trendy." By the late 1990s there was a wide recognition of the "gentrification" process within which the cultural industries were embroiled, as developers became increasingly aware of their value and adept at attracting them. The story is in fact more complex than "gentrification," with white cubes and espresso bars landing in the middle of deprived communities and cheap property. Areas such as Nottingham's Lace Market, Liverpool's Duke Street/Bold Street, Clerkenwell in North London, and Manchester's Northern Quarter, as well as areas such as Chicago's Wicker Park (Lloyd 2006), all had longer histories of artist and cultural business involvement in local social, environmental, and political issues. If gentrification implies a process of residential substitution of lower by higher income socioeconomic groups, then this was also about the substitution of cultural production by new kinds of up-market residential development and its lifestyle consumption accoutrements. The substitution of spaces of production by spaces of consumption, identified by Sharon Zukin (1995), cuts across creative, as well as older, manufacture-based, industries. This is exemplified by the recent adoption by many local authorities of Florida's "creative class" idea and their required cultural infrastructure; "creative industries" give way quite rapidly to creative consumption (Oakley 2009).

For example, in the late 1980s Manchester's Northern Quarter emerged as a cheap, atmospheric location for small-scale cultural businesses. The local community body, the Northern Quarter Association (NQA), was made up of local residents, shopkeepers, and an overwhelming number of "cultural" people located in the area. They got deeply involved in local community politics and the day-to-day negotiations with the city council concerning the management of the area (which also contained many "problem" social agencies, such as homelessness-, drug-, and alcohol-related services). Local creative businesses were well aware of and concerned about the impact of their presence on the qualities of place that had attracted them in the first place. There was a strong sense of historical recovery, of old buildings, dug out from years of abusive "renovation," but also of older working-class traditions in the area—pet shops, weaving, political radicalism, popular department stores, and so on. Creative businesses had a high participation rate in the local association and were generally tolerant of the different users of the area (homeless people, alcoholics and needle exchange, cheap shops and unruly pubs, etc.). The ability of the NQA to articulate the concerns and aspirations of local residents and businesses related to the tradition of community politics inherited from the 1970s and 1980s, the familiarity many cultural businesses had with bureaucratic procedures (funding applications, planning permission, etc.), and the confidence in challenging official definitions common among highly educated groups.[7] Although small-scale development has happened through the 1990s, notably through the local company Urban Splash, it was not until 2002–3 that the city mobilized developers around a concerted plan for the area. The first step in this process was the abolition of the NQA as a "redundant" organization and its replacement by informal consultation. The NQA had a legal constitution and a legal requirement to report through minutes and annual meetings; informal, nonbureaucratic consultation was invisible and had no reporting requirements. Nor did it ever do so, except through glossy newsheets.

We would argue that, certainly up to the end of the 1990s, the cultural and creative industries were not simply agents of gentrification but could be an important driver for a progressive reinvention of urbanity (Corijn 2009). "Gentrification" in UK city centers most frequently occurred when large-scale development capital, through the backing of city council regeneration plans, took control of "trendy" areas and began to substitute (relatively) low-rent production spaces for higher yield consumption, residential or "clean" or "premium" office space. The abolition of the NQA as precondition for this process merely illustrated in fairly stark terms what happened in many other locales. Nevertheless, we also need to note that the nature of this local "development capital" was very much inflected by its closer engagement with culture-led regeneration in the 1980s and 1990s.

In Manchester (as elsewhere) it was through involvement and negotiations around culture-led urban regeneration that the creative industries became involved in policy issues, rather than a concern for economic development per se. Research by Manchester Institute for Popular Culture (MIPC) in the mid to late 1990s suggested that it was with the quality of the urban realm and its wider image or brand that the creative industries were most concerned, and which they also saw as their main input into policy—

if anyone would listen.[8] Indeed, it was commonly felt in this period that the most successful forms of local support were those public events ("In The City" music convention, Mardi Gras, various film, fashion, design and theater festivals, etc.) that showcased both the local cultural industries and promoted the wider image of the city at the same time. In many respects the official and unofficial cultures of the city operated in two distinct spheres (O'Connor and Wynne 1996; Haslam 1999). The IRA bombing of the city in 1996 saw a major "rapprochement" between the established urban elites, increasingly concerned to promote the city (especially through two failed but catalytic Olympic bids) and newer voices from the recent ("rave" or "madchester") music, bar, and club scenes in the city. Not only had the increasing occupation of the city center through creative workspace and nighttime usage transformed the symbolic landscape of the city, initially for participating groups; many music-industry actors and associates moved into small-scale property development. Owners of the first gay bar, Mantos, and another "café-bar" leader, Atlas; the manager of the band Simply Red; a small-scale music merchandise entrepreneur, Tom Bloxham (of Urban Splash)—all these moved into property development, some of them now operating on a national scale. Equally, many of those involved in this field, most notably Tony Wilson, director of Factory Records (and of *24 Hour Party People* fame), went on to occupy major policy positions within the city. A key moment in this shift from the margins to the center within Manchester's elite growth coalition was the formation, by some of those just listed, of a pressure group to reject the official city marketing slogan as "outdated" and inappropriate for a city of Manchester's cultural aspirations (Ward 2000). A new post-rave urban growth coalition emerged that valorized entrepreneurialism, culture, and the creative industries as key to the future of the city—this at the same time when New Labour was making similar claims at the national level (Peck and Ward 2002; Ward 2000; Hetherington 2007; O'Connor 2007b).

We would argue that rather than simply "sector development," in the United Kingdom, creative industry policy has mostly been a kind of urban policy—something implied by the "creative city" theses of Charles Landry (2000) and Richard Florida (2005), though underdeveloped (cf. Scott 2001). In terms of an emerging urban cultural and "creative industries" policy narrative there was certainly a symbiosis between the city council and local elites and the creative sector in terms of image, with the latter clearly benefiting from a locational "brand." This is even more marked among those with property interests in the city. This symbiosis resulted less in calls for specific economic development support to creative businesses, and more a "recognition" of their contribution to this local identity and image. "Recognition" here might be understood not just as acknowledgment of their contribution to the "brand" but also of their legitimate input into policy development through a range of formal and informal forums and spaces.[9]

At the same time, this symbiosis relates to the fact that these businesses are deeply embedded in, and committed to, local social, cultural, and urban contexts. It forms the contexts for their practice as individuals, businesses, and networkers (O'Connor 1998; Drake 2003; Shorthose 2004; Banks 2007). This was a much more ambiguous relationship as it could cut across the interest in development gain through property prices that drove the new city center growth coalition. Thus, the concerns of many in the creative industries could be swept aside by more powerful local developers now able to deploy the language of creative industries much more effectively than previously. In the Northern Quarter example, although in the 1990s the NQA could use the new language of culture-led regeneration in the face of old style planning regimes, in 2002–3 they were positioned as local, vested interests opposed to the requirements of a citywide creative industries strategy.

As previously discussed, who might claim to speak for, to represent, the creative industries was a key issue in this new policy field; part of this claim was also about the power to construct a new narrative for the city. In mid-to-late 1990s Manchester the consensus around the creative agenda hid many ambiguities. The activities of CIDS, consequently, tended to be as much about mobilizing this wider identification with and commitment to the city as it was about business services or marketing strategies. It thus inherited those ambiguities.

THE CREATIVE INDUSTRIES DEVELOPMENT SERVICE

Commissioned research in 1997–99[10] in Manchester suggested that some sort of agency be established to act as intermediary between the city's economic and business development infrastructure and the creative industries sector. In response, CIDS was established,[11] envisaged as having four main functions: first, to develop, commission, and deliver bespoke businesses services to the sector—based on the recognition that the delivery and content of these services required a specific expertise and a reputation (or "track record") not available to mainstream organizations; second, to provide information and strategic intelligence about the sector to the city and to provide relevant information (services, grants, other opportunities) to the sector itself; third, to build partnership and ensure coherent delivery from the many agencies that were somehow involved in supporting the creative sector; and forth—the most difficult function involved a representative role—helping the sector to identify and articulate its particular requirements and giving voice to these concerns needs within the wider policy forums in the city.

Issues of voice and representation concern positions of power and legitimacy within the economic and urban policy field. The entry of the creative industries introduced much uncertainty into this field, for the reasons indicated already. If CIDS claimed a representative function, it did so not in a formal institutional or electoral sense, but rather by claiming both to speak the language of the creative sector and also to be able to translate this into the more formal language of economic development policy. Its intermediary function was thus established not simply by statistical or analytical knowledge but by the insider knowledge and relations of trust implied by this "speaking its language." *Language* here refers not only to very different ways of understanding and doing creative business; it also refers to the wider context of these creative businesses involved in a matrix of economic, social, cultural, and ethical imperatives. Part of intermediation was therefore "translation," and it hid a certain ambition: that CIDS would present the wider social and cultural (including urbanistic) concerns of the sector within the formal language of economic development to open the field of economic development to a recognition of the legitimacy of these wider social and cultural questions. It was here that the question of "cultural" or "economic" policy was so crucial. CIDS claimed a legitimacy denied the publicly funded cultural agencies in that it was ostensibly concerned with questions of commerce and economics. Only in this way could it sit down at the economic development policy table. At the same time, it tried to introduce a wider sociocultural agenda into this field—an attempt explicit in related notions of the "creative city," for example. It was a strategy that seemed feasible in a situation where economics and culture seemed much more open to each other. However, it was a also strategy beset by the danger, not just of losing the trust of both sector and policy agents ("falling between two stools"), but also by the latter discovering alternative languages for the creative sector, more directly amenable to mainstream economic development. Both of these occurred, as we shall see.

From the beginning, charged with the energy of the new creative industry agenda, CIDS tried to develop ways of thinking and acting differently from the mainstream public policy field. The first director was very clear that it was not about mechanically putting resources together but thinking critically about local creative industries and what they want. CIDS became somewhat unconventional, not trying to impose something from the top down but rather to be responsive, working as an interface between the creative sector and the local development infrastructure.[12]

It attempted to operate through the identification and promotion of networks within the sector, seeking "representative" voices. Such informal identifications were a matter of judgment based on track record, reputation, and the ability to express their needs and concerns in a way that spoke for the (sub)sector as a whole rather than their particular business. In this, CIDS was operating very much like creative businesses themselves, working through networks of reputation and trust. It then funded these individuals, or sometimes already existing networks, to develop a more inclusive network organization that could give voice to the needs of the sector and with whom CIDS could then work deliver policy initiatives in a convincing way. These included Red Eye (photographers), PANDA (performance), Manchester Fashion Network, Manchester Music Network, and M62 (computer games), among others. CIDS itself worked within a Manchester-based network of organizations and across the Northwest with other similar creative industries agencies, most notably ACME in Liverpool and the Arts Council-funded audiovisual agency North-West Vision (NWV).

Against initial expectations it was not through its specialized services (which it tended increasingly to subcontract and signpost rather than deliver) but through its day-to-day networking activities that it developed its range of contacts and a certain degree of trust: a reputation based not (primarily) on the formal knowledge provided by consultation reports but on contacts and interpersonal relationships. CIDS managed to build up a certain credibility as an advocate for the sector, one that spoke its language; it tried to ensure its intermediary role by embedding itself in the local cultural and social "scene." CIDS was located in a prominent (bright pink) creative industries building located in the heart of the Northern Quarter. Its staff were recruited mostly from outside business support personnel and tended to be close to (if not participants in) creative production. CIDS was therefore both "physically" and "socially" part of the creative cluster.

But this "lock-in" (or "going native") was also the source of tension on a day-to-day basis. CIDS was always hampered by its short-term funding structure (and it is not alone in this), whereby it could never fully guarantee the continuation of its different services. This uncertainty could undermine trust. Equally, while talking the language of the sector, CIDS also had responsibilities for monitoring and evaluation—basic and often clumsy metrics ("box-ticking") about business activities and impacts, which caused great friction between it and creative business. It was a bureaucratic process but it also undermined the claim of CIDS to be speaking a responsive language. If CIDS acknowledged that the development of creative industry networks is a process mixing cultural, social, and economic functions in ways difficult to separate, CIDS also knew that public intervention brings with it a different language, one that demanded a set of clear, usually economic, outcomes. Despite the attempts to reduce this to a (usually painful) lip service, monitoring and evaluation inevitably brought a gradual institutionalization and an accommodation with the overarching language of the existing policy world ("economic growth," "social inclusion," etc.).

Equally, the expectations raised by the organization within the sector were almost always too high for the level of resources commanded by CIDS, leading to frequent disenchantment. The establishment of trust by CIDS staff with the local businesses was often based on an expressed "commitment" to the creative industries sector, which, despite its personal sincerity, could not always be delivered by an organization constantly at the mercy of policy shifts and new funding regimes. Indeed, speaking the dual languages of creative sector and public policy caused much disorientation within CIDS staff. Personal commitment and sharing a "language" with the creative industries in the end could hardly make up for the lack of resources.

These tensions were felt within the networks supported by CIDS. These networks, many of which had been purely informal, had to be formalized to some degree to allow CIDS to fund them, and so as to develop a more systematic analysis of their needs (member surveys, for example) for policy purposes. This often undermined the more organic aspects of these networks. Many felt that these more targeted formal social network initiatives, as compared to practitioner-led social networks, focus less on personal issues and more on instrumental functions such as funding or organizing trade events. There were real worries about "being institutionalized" and anxieties as to the extent to which such formalization of communication is useful in forging meaningful interaction among creative businesses. And despite the ambition to be part of the sector, there was always an inevitable gap between (relatively secure) publicly funded agency and precarious business. In this context it was very easy for the relationship to move from one of trust to one of exploitation and cynicism.

CREATIVE INDUSTRIES AS URBAN POLICY

Although there are certain structural tensions between CIDS's necessarily more formal network initiatives and the informal or "organic" business initiated networks, there seemed real value in the development of networking infrastructures in local creative industries as part of the functioning of an intermediary organization. These reasons do not just concern the efficiency and customized nature of the business support that they facilitated. Creative industry clusters and networks have social and cultural as well as economic dimensions; they are embedded in the fabric of the city. The manner of this embedding does not simply mean the economic gains of "untraded interdependencies," skills pools, and business networking but is also about a reflective engagement with the cultural and social and environmental context of these localities. As such, CIDS's remit inevitably extended beyond business support to policy debates across a wide front—from city marketing to business support, to skills development and knowledge transfer; from protecting "independent business" in the city to reporting on development tensions in the Northern Quarter; from commenting on DCMS policy documents to co-coordinating responses to anxieties about the BBC's planned relocation to MediaCity. As such, CIDS attempted to move beyond sector-specific business support to a more general policy agenda intended to reflect the creative industries' identification with and support for the brand of the city—something the city mostly welcomed—and their engagement with the social, cultural, and urban contexts of their day-to-day activities—something that was much more contentious.

As we previously noted, the creative industries agenda has been set in a local economic development context, but one predicated on wider notions of culture and creativity. The diverse group of businesses and freelancers and public and private agents covered in the term "creative industries" very much bought into this. The creative industries certainly shared a narrative of "urban regeneration" in Manchester, one that was particularly inflected by the distinct Manchester brand of popular music and culture (O'Connor 2004). This was a relationship formally cemented by the appointment of Peter Saville, of Factory Records fame,[13] as "creative director" for the city (O'Connor 2007b). But this shared narrative was subject to ongoing negotiation and could easily be abused—as in the well-established model where property developers emerge as the real winners in "culture-led regeneration." What CIDS attempted to do was try to create a "shared narrative" between a public agency and the creative industries. Only in this way, it was felt, could long-term trust be built and the intermediary function of channeling "constituency needs" into public policy be performed. We have seen some clear limits to this, in terms of what more formal public policy language and process might allow, and what difficulties it imposes on such trust building. If the dual-facing intermediary function placed strains on the CIDS claim to speak the language of the sector, it did so also within the policy field. Though its brand value as a trusted agency was recognized,[14] its unorthodox approach and (relatively) disruptive language positioned it as somewhat of a maverick organization.

Therefore, a more telling limit was reached as the language of economic development, increasingly prevalent within the creative industries discourse, allowed a displacement of these wider sociocultural and urbanistic concerns. In a manner parallel to the trajectory of the NQA, as the field of creative industries policy became more established and its language better known, the specific value of CIDS as intermediary could more easily be marginalized. That the discourse of "creative industries" introduced an increasingly economic rationale into debates around culture-led regeneration and the creative

city, as well as within cultural policy as a whole, is little disputed as such—merely the legitimacy of such a shift (cf. Cunningham 2004; 2008; Hesmondhalgh and Pratt 2005; Banks and O'Connor 2009). Summarizing a more complex argument, we suggest that though the notion of creative industries does not have to be interpreted in this way (and see Cunningham 2009 for a global survey), in the UK context at least, it has allowed a de-complexification of policy, dropping those considerations external to purely economic outputs (see O'Connor 2009). However, these economic outputs are now much more easily couched in the language of "creativity," and hence linked to the culture-led narrative of the city.

This process can be seen in the developments around MediaCity in Salford Quays. The announcement of a relocation of BBC production facilities from London to Manchester raised the profile of the creative industries sector overnight. Not only was this a prestigious image coup for the city, underlining the success of its rebranding, but it would also represent new jobs and a substantial property investment. That it eventually located in Salford, a separate adjacent local authority, is not something we can discuss here. However, the raised profile of, and stakes involved in, leading the creative sector meant that CIDS was marginalized within a short period of time. That it made some fatal mistakes in the game of political maneuvers is a commonly held view, as we discovered in recent interviews. But that the game suddenly involved more powerful and adept players is not in dispute either. The North West Regional Development Agency (NWDA) strongly supported the relocation and used it as the centerpiece of its Digital and Creative Industries Sector Development Plan (NWDA 2008)—which was to inform all its funding for creative industries development in the near future. Key to the strategy was MediaCity as the attractor for major foreign direct investment and the creation of Europe's "second largest media and digital cluster" (after London).

Although in the relatively open situation of the late 1990s local and regional bodies were willing to see CIDS as a legitimate voice for an unknown and ill-defined sector, the growing visibility of creative-sector leadership were such that it could not be left to a maverick agency such as CIDS. Or rather, it responded to the shift in language, from a cultural to an economic policy emphasis, too late. Echoing the DCMS's inflation of statistics by the inclusion of software development (Garnham 2005), the NWDA plan has the thirteen original DCMS sectors and also information and communications technology (ICT)—which includes a huge number of companies involved in cable and telecommunications. It is little wonder that, according to NWDA, only thirty companies in the Northwest sector employ more than 250 people and these are "mostly digital" (i.e., ICT). Equally, the inclusive "creative industries" has been decomposed into "media," "digital," and "the rest," with the first pair being seen as the key economic drivers. The language of the plan gives no quarter to any cultural policy concerns, and no consideration of the urbanistic context of the creative industries. In addition, rather than delivering policy through a complex network of subregional organizations that could respond to local contexts, the head of creative industries at NWDA decided—in a field defined by its complex constituencies and policy spheres—to "simplify" delivery through one regional organization. In this context an agency that had been dedicated to supporting television and film across the region, even though it lacked all knowledge of the complex on the ground delivery associated with CIDS and other local agencies, was well-placed to succeed. North West Vision now leads all creative industry support across the region. At this point, the CIDS director "threw in the towel" and the organization effectively ceased at the end of 2008.

The consultation process in response to the plan, which was led by CIDS, had been a rough ride. The various local authorities objected that all resources were going into one central node; the network of local agencies objected to subcontracting to one, inexperienced regional agency (though this was left as ambiguous for some time); and they also objected to the exclusive focus on "digital" when sectors such as music had long been associated with the region. Those in the creative sector objected that this strategy ignored the complex urban ecology out of which Manchester's creative industries had emerged in the first place. That is, as with the process of "gentrification," the revived image of the city would be used to build a high-rent media park aimed at large-scale international business at the expense, it was claimed, of the local creative sector. However, the global financial crisis has since undermined the extravagant claims for the attraction of major international media and digital firms—especially in a city with no direct flights to the U.S. West Coast or the Far East. Recent interviews suggest that in the downturn, small-scale creative companies are thriving, with the UK government recently launching a scheme to bring them back into city centers to occupy empty property (Booth 2009).

CONCLUSION

In this article, we have tried to argue that creative industries policy should be conceived as a kind of urban policy. Creative industries benefit from and contribute to the image of the city, but they are also embedded in its social and cultural life. Creative industries tended to actively share narratives of urban regeneration, local identity and the "creative city." These were valued more than initiatives around training, business support and advice, and so on, of which they were and are skeptical. CIDS emerged in this context to enhance support for the sector by giving it a coherent channel through which to voice its needs. It was able to identify and translate this voice into formal policy due to a trust built around a shared commitment or identification. This shared identification inevitably brought social and cultural concerns into the field of local economic policy. CIDS attempted to manage and voice these concerns within the relatively open space of the creative industries agenda, extending its remit across a wide range of local policy issues.

We could say that CIDS had a hidden cultural political agenda: that it would represent the needs of the sector at policy level not just to reengineer its business services accordingly but to represent the social and cultural aspirations of the creative industries sector at the wider city level. "Support for creative businesses" then becomes not just a matter of bespoke business services, or their delivery in the appropriate language and context, but also about recognizing their needs and contributions beyond economics; it's a recognition that often cuts against a wider economic logic to include claims to cultural integrity and social justice.

The attempt was not to oppose some "economic logic" but to act on the implications of the creative industries agenda as expressed in the terminology of local clusters, creative ecology, soft infrastructure, creative field, and so on. All these implied an expanded economic logic, and one that necessarily involved social and cultural considerations. This was the promise of the cultural and creative agenda to many of its proponents. But in reality such complex policy vision and the subtle knowledge that this might demand were always lacking in local authorities. The splits between economic policy, urban planning, and cultural policy—perhaps mirrored at UK national level by the imperatives of the Treasury as the framework against which policy impact must be measured—were too deep to be traversed by this agenda.

Thus, supporting clusters of creative industries in areas that do not seem to promise immediate economic benefits makes little sense to economic development agencies; the logic suggests that they pursue global media companies instead. In such circumstances, the difficult balance between the language of public policy and that of industry, and between economic and sociocultural value, can disappear under the purely economic language of "creative economy," "innovation," and related discourses that see the language of "culture" as a remnant of cultural policies past (Cunningham 2004).

CIDS has been caught up in the success of the creative industry argument; as it has moved ever more center stage in local policymaking in Manchester (and elsewhere), the power to decide on policy and to "speak" on behalf of the sector has shifted further and further away. "Media" and "digital" industries, now economic priorities for the Manchester region, have long been allocated to other agencies. It is these who now sit at the policy table speaking for those creative industry sectors that really matter. The generic creative industry sector, which CIDS attempted to represent, is now seen as more peripheral, and the social and cultural concerns with which it arrived in the 1990s are downplayed as "lifestyle" rather than serious businesses. The recession may have changed this.

NOTES

1. An interesting attempt was that of Taylor, Evans, and Frasier (1996), bringing Williams's national-level "structure of feeling" down to local city-region level—in this case, contrasting Sheffield and Manchester.

2. This is obviously linked to different political systems and their structuring of local and national—for example, unlike the United Kingdom, local power bases in France can launch national careers.

3. See Oakley (2009) for a retrospective discussion.

4. The emergence of new cultural policies, cultural industries policies, and creative city policies in places such as Sheffield, Birmingham, Greater Manchester, Bristol, and Huddersfield related very much to specific individuals—and often dissipated with their departure.

5. The list is extensive: Birmingham's custard factory; Helsinki's cable factory; Marseilles's cigarette factory; Factory 798 in Beijing; Amsterdam's Westergasfabriek and Witte Dame, Eindhoven; London's Truman Brewery and Berlin's Bruerei; Moscow's winery; and so on.

6. http://www.culture.gov.uk/what_we_do/creative_industries/3275.aspx

7. In particular, the history of Manchester's Hulme area, associated with the punk and postpunk era of the city, had seen high levels of oppositional engagement with the political, legal, and bureaucratic processes of urban regeneration (see Haslam 1999; 2005; Dickinson 1997).

8. MIPC was established in 1993 at Manchester Metropolitan University (Redhead, 2004). Justin O'Connor was director 1995–2006, and then it was closed. Its work on local urban cultures included a conference on the "night-time economy" in 1994; a consultancy on the Northern Quarter for Manchester City Council 1995–96; three UK Economic and Social Research Council Projects (cf. Wynne and O'Connor 1998; O'Connor 1998; Brown, Cohen, and O'Connor 2000; Banks, O'Connor, and C. Raffo 2000; Banks 2007); and a large-scale study, "Cultural Production in Manchester," which provided the research basis for CIDS. This document became publicly unavailable after the closure of MIPC.

9. We might say it echoes the way contemporary games companies involve users in brand development, and are required to acknowledge this through respecting the rules of engagement of this user input (Banks and Humphreys 2008).

10. See note 8.

11. Funded mainly via European Structural Fund money through Manchester City Council, then via funds from the North West Regional Development Agency.

12. This section on CIDS is based on interviews conducted by the authors. O'Connor was co-chair of CIDS between 2000 and 2006. Xin Gu conducted extensive interviews as part of her PhD thesis: "Social Networks in Cultural Industries: Fashion, New Media and Network Development Policy in Manchester," Manchester Metropolitan University, 2008.

13. Saville created the distinctive image of Factory Records and its club, The Hacienda (see Haslam 1999).

14. For example, CIDS became a "must-see" organization whenever cultural and creative industries delegations came to visit the city.

REFERENCES

Anheier, H., and R. Isar. 2008. *The cultural economy, culture and globalisation 2*. London: Sage.

Banks, M. 2007. *The politics of cultural work*. Basingstoke: Palgrave.

Banks, J., and S. Humphreys. 2008. The labour of user co-creators: Emergent social network markets? *Convergence* 14(4):401–18.

Banks, M., and D. Hesmondhalgh. 2009. Looking for work in creative industries policy. *International Journal of Cultural Policy* 15:415–30.

Banks, M., and J. O'Connor. 2009. Introduction to "Ten Years After." *International Journal of Cultural Policy* 15:365–73.

Banks, M., A. Lovatt, J. O'Connor, and C. Raffo. 2000. Risk and trust in the cultural industries. *Geoforum* 31:453–464.

Bell, D., and M. Jayne, eds. 2004. *City of quarters: Urban villages in the contemporary city*. Aldershot, UK: Ashgate.

Bianchini, F. 1987. GLC R.I.P. 1981–1986. *New Formations* 1:103–17.

Bianchini, F., and M. Parkinson, eds. 1993. *Cultural policy and urban regeneration: The West European experience*. Manchester: Manchester University Press.

Bilton, C. 2007. *Management and creativity: From creative industries to creative management*. Oxford: Blackwell.

Booth, R. 2009. Ministers outline emergency measures to prevent the recession creating ghost towns across the UK. *Guardian*, April 14, 2009. http://www.guardian.co.uk/business/2009/apr/14/government-high-street-shops-grants (accessed April 14, 2009).

Brown, A., S. Cohen, and J. O'Connor. 2000. Local music policies within a global music industry: Cultural quarters in Manchester and Sheffield. *Geoforum* 31(4):437–51.

Corijn, E. 2009. Urbanity as a political project: Towards post-national European cities. In *Creative economies, creative cities: Asian-European perspectives,* eds. L. Kong and J. O'Connor, 197–206. Dordrecht, The Netherlands: Springer.

Cunningham, S. 2004. Creative industries after cultural policy. *International Journal of Cultural Studies* 7:105–15.

Cunningham, S., J. Banks, and J. Potts. 2008. Cultural economy: The shape of the field. In *The cultural economy,* eds. H. Anheier and R. Isar, 15–26. London: Sage.

Cunningham, S. 2009. Trojan horse or Rorschach blot? Creative industries discourse around the world. *International Journal of Cultural Policy* 15:375–86.

DCMS. 1998. *Creative industries mapping document*. London: Author.

Dickinson, R. 1997. *Imprinting the sticks: The alternative press beyond London*. Aldershot, UK: Ashgate.

Drake, G. 2003. This place gives me space: Place and creativity in the creative industries. *Geoforum* 34(4):511–24.

Florida, R. 2005. *Cities and the creative class*. New York: Routledge.

Garnham, N. 1983/1998. Concepts of culture: Public policy in the cultural industries. *Cultural Studies* 1(1): 23–37.

Garnham, N. 2005. An analysis of the implications of the 'creative industries' approach to arts and media policy making in the United Kingdom. *International Journal of Cultural Policy* 11:15–29.

Granovetter, M. 1985. Economic action and social structure: The problem of embeddedness. *American Journal of Sociology* 91:481–510.

Hall, P. 2000. Creative cities and economic development. *Urban Studies* 37:639–49.

Hall, S., and M. Jacques.1989. *New times*. London: Lawrence and Wishart.

Hannigan, J. 1998. *Fantasy city: Pleasure and profit in the postmodern metropolis*. London: Routledge

Hartley, J. 2005. Creative industries. In *Creative industries*, ed. J. Hartley, 1–40. Oxford, UK: Blackwell.

Harvey, D. 1989. *The condition of postmodernity*. Oxford: Basil Blackwell.

Haslam, D. 1999. *Manchester, England: The story of the pop cult city*. London: Fourth Estate.

Haslam, D. 2005. *Not Abba: The real story of the 1970s*. London: Fourth Estate.

Hesmondhalgh, D. 2007. *The cultural industries*. London: Sage.

Hesmondhalgh, D., and A. Pratt. 2005. Cultural industries and cultural policy. *International Journal of Cultural Policy* 11:1–14.

Hetherington, K. 2007. Manchester's Urbis: Urban regeneration, museums and symbolic economies. *Cultural Studies* 21:630–649.

Hewison, R. 1997. *Culture and consensus*. London: Methuen.

Howkins, J. 2001. *The creative economy: How people make money from ideas*. London: Penguin.

Hutton, W. 1995. *The state we're in*. London: Cape.

King, R. 2006. *Detonation: Rebirth of a city*. Warrington, England: Clear.

Kong, L., C. Gibson, L. M. Khoo, and A. L. Semple. 2006. Knowledges of the creative economy: Towards a relational geography of diffusion and adaptation in Asia. *Asia Pacific Viewpoint* 47:173–94.

Kong, L., and J. O'Connor, eds. 2009. *Creative economies, creative cities: Asian-European perspectives*. Dordrecht, The Netherlands: Springer.

Landry, C. 2000. *The creative city*. London: Comedia.

Lloyd, R. 2006. *Neo-Bohemia: Art and commerce in the postindustrial city*. London: Routledge.

Manchester City Council, eds. 2002. *The Mancunian way*. Manchester: Clinamen Press.

———. 2004. *Manchester: Shaping the city*. London: RIBA.

Markusen, A. 1996. Sticky places in slippery space: A typology of industrial districts. *Economic Geography* 72:93–313.

Marshall, A. 1890. *Principles of economics*. London: Macmillan.

Martin, R., and P. Sunley. 2003. Deconstructing clusters: Chaotic concept or policy panacea? *Journal of Economic Geography* 3:5–35.

McRobbie, A. 1998. *British fashion design: Rag trade or image industry?* London: Routledge.

Mommaas, H. 2004. Cultural clusters and the post-industrial city: Towards the re-apping of urban cultural policy. *Urban Studies* 41(3):507–32.

———. 2009. Spaces of culture and economy: Mapping the cultural-creative cluster landscape. In *Creative economies, creative cities: Asian-European perspectives*, eds. L. Kong and J. O'Connor, pp. 45–60. Dordrecht, The Netherlands: Springer.

Montgomery, J. 2007. *The new wealth of cities: City dynamics and the fifth wave*. Aldershot, UK: Ashgate.

Myerscough, J. 1988. *The economic importance of the arts in Britain*. London: Policy Studies Institute.

North West Development Agency. 2008. *Digital and creative industries: Sector strategy action plan*. Warrington, England: North Western Development Agency. http://www.nwda.co.uk/areas-of-work/business/key-sectors/digital-and-creative.aspx (accessed March 30, 2009).

Oakley, K. 2009. Getting out of place: The mobile creative class takes on the local. A UK perspective on the creative class. In *Creative economies, creative cities: Asian-European perspectives*, eds. L. Kong and J. O'Connor, 121–34. Dordrecht, The Netherlands: Springer.

O'Connor, J. 1998. New cultural intermediaries and the entrepreneurial city. In *The entrepreneurial city: Geographies of politics, regime and representation*, eds. T. Hall and P. Hubbard, 225–40. Chichester, UK: Wiley.

———. 2000. The definition of the cultural industries. *European Journal of Arts Education* 2:15–27.

———. 2004. A special kind of city knowledge: Innovative clusters, tacit knowledge and the 'creative city'. *Media International Australia* 112:131–49.

———. 2007a. *The cultural and creative industries. A review of the literature*. London: Creative Partnerships/Arts Council England.

———. 2007b. Manchester: The original modern city. *The Yorkshire and Humber Regional Review* Special edition:13–15.

———. 2009. Creative industries: A new direction. *International Journal of Cultural Policy* 15:387–402.

O'Connor, J., and D. Wynne. 1996. Left loafing: City cultures and postmodern lifestyles. In *From the margins to the centre: Cultural production and consumption in the post-industrial city*, ed. J. O'Connor and D. Wynne, 49–90. Aldershot, UK: Ashgate.

Parrish, D. 2005. *T-Shirts and suits—A guide to the business of creativity*. Liverpool: ACME.

Peck, J., and K. Ward, eds. 2002. *City of revolution. Restructuring Manchester*. Manchester: Manchester University Press.

Peck, J., and M. Emmerich. 1992. Recession, restructuring... and recession again: The transformation of the Greater Manchester Labour Market. *Manchester Geographer* 13:19–46.

Piore, M., and C. Sabel. 1984. *The second industrial divide: Possibilities for prosperity*. New York: Basic.

Porter, M. E. 1998. Clusters and the new economics of competitiveness. *Harvard Business Review* 76:77–90.

Porter, C. 2000. Location, competition and economic development: Local clusters in a global economy. *Economic Development Quarterly* 14:15–34.

Powell, W. 1990. Neither market nor hierarchy: Network forms of organization. *Research in Organizational Behaviour* 12:295–336.

Pratt, A. 1997. *The cultural industries sector: Its definition and character from secondary sources on employment and trade, Britain 1984–91*. Working paper on Environmental and Spatial Analysis, No. 41. London School of Economics and Political Science, London, England.

Pratt, A. 2000. New media, the new economy and new spaces. *Geoforum* 31:425–36.

———. 2002. Hot jobs in cool places. The material cultures of new media product spaces: The case of South of the Market, San Francisco. *Information, Communication and Society* 5:27–50.

———. 2004. The cultural economy: A call for spatialised 'production of culture' perspectives. *International Journal of Cultural Studies* 7:117–28.

———. 2009. Policy transfer and the field of cultural and creative industries; What can be learned from Europe? In *Creative economies, creative cities: Asian-European perspectives*, eds. L. Kong and J. O'Connor, 9–24. Dordrecht, The Netherlands: Springer.

Pratt, A., and P. Jeffcut. 2009. Creativity, innovation and the cultural economy: Snake oil for the 21st century? In *Creativity, innovation and the cultural economy*, eds. A. Pratt and P. Jeffcut, 1–19. London: Routledge.

Raffo, C., M. Banks, A. Lovatt, and J. O'Connor. 2000. Attitudes to formal business training and learning amongst entrepreneurs in the cultural industries: Situated business learning through 'doing with others'. *British Journal of Education and Work* 13(2):215–30.

Redhead, S. 2004. Creative modernity: The new cultural state. *Media International Australia* 112:9–28.

Roodhouse. S. 2006. *Cultural quarters: Principles and practice*. Bristol, UK: Intellect.

Schlesinger, P. 2007. Creativity: From discourse to doctrine. *Screen* 48(3):377–87.

Scott, A. J. 2000. *The cultural economy of cities*. London: Sage.

Scott, A. J. 2001. Capitalism, cities and the production of symbolic forms. *Transactions of the Institute of British Geographers* 26:11–23.

Scott, A. J. 2004. Cultural products industries and urban economic development: Prospects for growth and market contestation in global context. *Urban Affairs Review* 39:461–90.

———. 2006. Entrepreneurship, innovation and industrial development: Geography and the creative field revisited. *Small Business Economics* 26:1–24.

Shorthose, J. 2004. Nottingham's de facto cultural quarter: The Lace Market, independents and a convivial ecology. In *City of quarters: Urban villages in the contemporary city*, eds. D. Bell and M. Jayne, 149–62. Aldershot, UK: Ashgate.

Storper, M. 1995. The resurgence of regional economies, ten years later: The region as a nexus of untraded interdependencies. *European Urban and Regional Studies* 2:191–221.

Taylor, I., K. Evans, and P. Fraser. 1996. *A tale of two cities: Global change, local feeling and everyday life in the North of England. A study in Manchester and Sheffield*. London: Routledge.

Ward, K. 2000. Front [sic] rentiers to rantiers: "Active entrepreneurs," "structural speculators" and the politics of marketing the city. *Urban Studies* 37:1093–1107.

———. 2003. Entrepreneurial urbanism, state structuring and civilizing 'new' East Manchester. *Area* 32(2):116–27.

Wynne, D., eds. 1992. *The culture industry: The arts in urban regeneration*. Aldershot, UK: Ashgate.

Wynne, D. and J. O'Connor. 1998. Consumption and the postmodern city. *Urban Studies* 35:841–864.

Zeitlin, J., and P. Hirst, eds. 1989. *Reversing industrial decline? Industrial structure and policy in Britain and her competitors*. Oxford, UK: Berg.

Zukin, S. 1982. *Loft-living: Culture and capital in urban change*. London: John Hopkins Press Ltd.

Zukin, S. 1991. *Landscapes of power: From Detroit to Disney World*. Berkeley: University of California Press.

Zukin, S. 1995. *The culture of cities*. Oxford: Blackwell.

Can Beijing become a Global Media Capital?

Angela Lin Huang

Beijing Research Centre for Science, Beijing Academy of Science and Technology, Beijing, China

INTRODUCTION: CHINESE MEDIA GOING GLOBAL

The Flowers of War (*jinling shisan chai*) was a 2011 Chinese film directed by the famous Chinese director Zhang Yimou, and starring the Hollywood actor Christian Bale. It was the most expensive film ever made in China, with a budget of 600 million RMB ($US94 million), and it had earned 592.5 million RMB ($US93 million) in box office receipts in the period from December 2011 to January 2012 alone. Despite its enormous commercial success in China, it failed to be shortlisted for Best Foreign Language Film at the 2012 Academy Awards, nor was it successful in its nomination at the Golden Globe Awards. For Zhang Yimou, this is the eighth time that he has sought but failed to receive an Oscar, and this can be seen as reflecting the continued inability for Chinese films to receive the international recognition commensurate with their domestic prominence.

A rapid process of commercialisation has taken place in China's film and television industries since the 1980s. Chinese media companies inspired by the success of *Crouching Tiger; Hidden Dragon* have been increasingly venturing into the international media market. Chinese filmmakers produce Chinese versions of blockbusters, by imitating the Hollywood generic style and incorporating Chinese cultural and historical themes. China Central Television (CCTV) has been expanding its broadcasting reach to countries in Europe, the Americas and Africa, with the aim of being a leading global broadcasting service.

The Chinese government is a key player in these media internationalisation campaigns. At the Third Plenary Session of the Seventeenth Central Committee in 2008, the Chinese President and the General Secretary of CCP, Hu Jintao, called for the development of top-ranking internationalised media that will spread Chinese images, voices, words and information across the globe. However even with the government's backing and financial support, China has still not been able to turn around its deficit in cultural trade. In 2010, about 260 domestic films were shown in Chinese cinemas and earned half of the total box office in China. The other half of the 2010 box office was accounted for by ten imported Hollywood blockbusters. Moreover, despite CCTV's rapid international expansion, CCTV never announces what its audience ratings actually are in overseas markets

Over the last three decades, Beijing, as China's cultural and political capital, has become a media center: directors, producers, actors and investors from Shanghai, Guangzhou, Hong Kong, Taiwan and Singapore have all been moving to find work and make deals (see Chen 2004). The Beijing municipal government aspires to develop the city as 'the oriental capital of film and television', according to Beijing's Twelfth Five-Year Plans released in November 2011. The aim is for the Central Business District (CBD) and Chaoyang District in the city's east to become the centre of a cluster of international media companies. But the question is widely asked as to how realistic are the aspirations of Beijing's authorities?

In exploring this question, this chapter makes use of Michael Curtin's concept of *media capitals* as providing an analytical framework for considering contemporary developments in the spatiality of Beijing's fast-changing media industries (Curtin 2003, 2007, 2010a, 2010b). In attempting to better understand the spatial dynamics of media industries and their propensity to cluster in certain geographical locations, Curtin has proposed that media capitals are

> centres of media activity that have specific logics of their own; sites of mediation; locations where complex forces and flows interact; meeting places where local specificity arises out of migration, interaction and exchange; places where things come together and ... consequently, where the generation and circulation of new mass culture forms become possible. (Curtin 2003: 203–205)

The chapter outlines Curtin's general theory of media capitals, and the interaction it proposes between economic logics of accumulation, trajectories of creative migration, and forces of socio-cultural variation, in the rise of media capitals outside of Hollywood.

In considering the case of Beijing, it is noted that in this chapter that the influence of local policies, both negative and positive, needs to be further identified in more depth than Curtin's approach tends to do, as Curtin focuses more on global market dynamics than on interventionist state policies. This has particular relevance to the study of Asian media capitals, as TV and film industries are typically strongly supported by national governments in the region, as they aim to develop a favourable climate for 'national champions' in the creative industries. The chapter also investigates the media environment that has formed, and is continuing to re-form in Beijing, and how participants, including state-run media organisations, private media companies and international media conglomerates are seeking out space and networks to develop their operations in the complex media ecology of Beijing. It draws extensively upon interviews undertaken with creative industries workers, company owners and producers undertaken in Beijing during October–November 2009.

THE CONCEPT OF MEDIA CAPITALS

Michael Curtin (2003, 2007, 2010a, 2010b) has proposed the concept of media capitals in order to explain the spatial dynamics of media industry agglomeration. The concept combines consideration of industrial infrastructure and economic dynamics (logic of accumulation), human capital formation (trajectories of creative migration), media and cultural policies and elements of cultural specificity (forces of social-cultural variation), in order to understand the rise of media capitals outside of the United States which can challenge the hegemony of 'Global Hollywood' (Miller et al. 2005).

Curtin's argument is framed by the proposition that the global media system today is no longer largely characterised by a one-way flow of US programming to the rest of the world. Rather, according to Curtin, the increasing volume and velocity of multi-directional media flows is emanating from the rise of media capitals once considered to be on the periphery of the world media system, such as Cairo, Mumbai and Hong Kong, as well as by US cities such as Miami, which largely interact with Spanish-speaking markets (Curtin 2007). Curtin maintains that media capitals are places where new mass culture forms containing local specificity are generated and exported, while media resources and talent come together, interact and engage in joint projects. Therefore, rather than seeing global media as merely a spatial expansion of US dominance – the 'Global Hollywood' thesis – the global media system is not only characterised by de-territorialisation and a global division of cultural labour, but also by the clustering of production activity in specific city-regions where there are particularly attractive opportunities for international investment (Storper 1997; Flew 2007). The argument is that the capacity and relative

global position of these emergent media capitals is advancing as they become hubs for various media flows conducive to the further accumulation of resources and new knowledge.

The concept of media capitals is a dynamic one, as the relative status of particular cities can rise or fall at different development stages. The concept of 'capital' in Curtin's framework is deliberately ambiguous, as a single word plays upon two meanings: a spatial location that is a centre of activity (the geographical definition), and a concentration of resources, reputation and talent (the economic definition). Michael Curtin understands the temporal dynamism and spatial complexity of media capitals within the global media as involving the interaction between three elements:

(a) *Logics of accumulation*: in this principle Curtin explains the centripetal tendencies of production and centrifugal tendencies of distribution from an economic perspective. The drive to maximise profits and accumulate capital is assisted by the development of ICTs and other technologies that revolutionise transport and communication, enabling modern enterprises to operate across national and regional boundaries to reduce production costs, expand their market reach and increase the speed of distribution. Curtin argues that, in this regard, media products are distinctive prototypes, rather than redundant batches of products with interchangeable parts, and that each product is unique.

(b) *Trajectories of creative migration*: creativity is a core resource to media industries and comes from pools of labour that continually develop new prototypes, motivated by a mix of desires for aesthetic innovation (intrinsic motivations) and identification of commercial opportunities (extrinsic motivations). As media capitals accumulate economic resources, media talent is attracted to these new production centres for the growing work opportunities, and the creative migration stimulated by new job opportunities enhances the attractiveness of the region to other media workers, which in turn drives the growth and expansion of media industries in the region.

(c) *Forces of socio-cultural variation*: media products can adapted to local traditions, resources and preferences, so as to be more readily accepted by audiences in particular geo-linguistic regions due to cultural proximity. Moreover, national and local institutions, ideologies, policies and cultures interact to create distinctive and complex variations between cities, regions and nations.

Curtin's work is largely derived from observations of the free market capitalist model in action, as seen in such case studies as Hong Kong, Mumbai and Hollywood. From such case studies, Curtin (2010a) has generalised that successful media capitals–such as Hollywood and Mumbai–have operated at arm's length from the state, catering to audience interests rather than to government *fiat*. Successful media capitals are thus seen to embrace commercial imperatives and respond rapidly to changing audience tastes, so that producers tend to avoid aligning themselves too closely to powerful patrons and official culture.

For these reasons, Curtin argues that, despite its global aspirations, Beijing is not a media capital but rather a national node of state policy and patronage (Curtin 2010a). He asserts that 'national political capitals tend not to emerge as media capitals, largely because modern governments seem incapable of resisting the temptation to tamper with independent media institutions' (Curtin 2010b: 266). Beijing in particular has created 'enclaves of "indigenous" production and topographies of national circulation' (Curtin 2010a), to the point where, in spite of its self-proclaimed aspirations to become the centre of Chinese media industries, the city's creative productivity is diminished by its thick political air and the oppressive official oversight. Accordingly, according to Curtin, Beijing will not be able to become the centre of Chinese commercial media sector, although it will still exercise its powerful influence over state-sanctioned media.

Nevertheless, the Chinese media industry has been described as a 'unity of contradictions'. Lee (2009) and Chan (1993, 2003) have observed that Chinese media has a double-edged nature, as it is both a state instrument for maintaining ideological control and a highly vibrant and marketised industry sector. The interweaving of market forces and government control fertilises a *quasi* market-oriented economy in China, in which state-owned media companies, private media companies and overseas media companies collaborate and compete under the supervision of the government. The Chinese government adopts strict measures to punish transgressors of its ideological norms,

and safeguards its authority in the 'forbidden zone' of ideological boundaries, which remain blurry to market players, especially non-Chinese players. This blurriness has been described as 'structured uncertainty' (Breznitz and Murphree 2011); in other words, there is a degree of uncertainty built-in to policy making and implementation. This uncertainty is linked to informal knowledge, of the rules of engagement, of the borders of expression. This uncertainty also enables media workers to bypass the forbidden zone by obtaining informal knowledge and building informal relationships with the government. As a result, there is a significant theoretical challenge in employing a concept such as Curtin's idea of media capitals, as it has been predominantly developed in a Western context. In order to better consider its relevance in non-Western contexts, and those where there are not clear boundaries between the state and the market, Beijing provides a suitable experimental field, as it is the city where there is the greatest concentration of media companies, talent and government organs in China.

THE LOGIC OF MEDIA ACCUMULATION IN BEIJING

Beijing, Shanghai and Guangdong Province are China's three most prominent 'media hotspots', or what Chinese scholars term 'media uplands' (*chuanmei gaodi*) (Tan and Wang 2009). Beijing prevails amongst these competing hotspots. Yu Guoming (2010) has evaluated media development in Chinese cities, taking into account five aspects – production; profits; consumption; advertising; and environment – and finds that Beijing is the leading media centre. In the TV sector, according to the Yearbook of Chinese Radio and Television (2009), Beijing, Shanghai and Guangdong Province accounted for more than half of the national TV market of China, with Beijing again assuming the leading position (see Table 6.1).

With respect to the film industry, according to the State Administration of Radio, Film & Television (SARFT),

TABLE 6.1

The gross income of TV sector in Beijing, Shanghai and Guangdong, 2006–2008

	2006		2007		2008	
	Sales of TV shows (million RMB)	Sales of TV drama (million RMB)	Sales of TV shows (million RMB)	Sales of TV drama (million RMB)	Sales of TV shows (million RMB)	Sales of TV drama (million RMB)
Beijing	502.48	395.86	710.63	438.35	1019.61	809.24
Shanghai	163.62	132.55	230.27	188.09	325.34	133.81
Guangdong	181.48	127.16	209.75	151.87	189.98	139.47
National	2079.44	1499.67	2368.85	1519.67	2420.66	1642.53

Source: Yearbook of Chinese Radio and Television 2009

total box office revenues exceeded 10 billion RMB in 2010 (SARFT 2011). Of the domestic films that were among the top ten best selling film of 2010 in China, five of the ten were produced or coproduced by domestic film companies located in Beijing, such as Huayi Brother Film Company and China Film Group: *After Shock* (*tangshan da dizhen*), *Let the bullets fly* (*rang zidan fei*), *If you are the one 2* (*feicheng wurao 2*), *Detective Dee and the mystery of the phantom flame* (*di renjie zhi tongtian diguo*), and *Ip man 2* (*ye wen 2*) (www.sina.com.cn, 2011). up. In addition, the China Film Group and Huaxia Film Distribution Company are the only two companies which are authorised to import and distribute foreign films, and both are located in Beijing.

Beijing not only prevails in mainland China, but is also attracting media personnel and investment from around the world, especially among the pan-Chinese community. Hui (2003) has observed that Hong Kong is experiencing an inevitable 'brain drain' as more multinational corporations move their Asia-Pacific development to Beijing and Shanghai, after China's accession to WTO and the small local market have severely hindered the further development of Hong Kong's TV and film industry. It has become imperative for Hong Kong producers to seek collaboration in mainland China. As the Hong Kong film producer, Wu Siyuan, has observed, Beijing has become China's film centre and, as a result, first releasing a film in Beijing is the most effective strategy in the Chinese market (cited in Zhang 2009).

Now, Beijing has the most media companies in China, especially large companies: the number of large film and TV companies in Beijing is double that of Shanghai, which is the second largest site (Liu 2009). In the TV sector, most provincial TV stations set up their offices and production base in Beijing, to be near the Beijing-based CCTV and BTV. In the production sector, there are 129 registered TV drama production companies and organisations in China, and almost a third – 41 – are in Beijing. CCTV is the biggest state-owned organisation in TV drama production, and was authorised by SARFT to produce 34 TV drama series (967 episodes) in 2008, while the Beijing-based CITVC and Beijing Zhongbei TV Art Centre Co. are also very important state-owned TV production organisations.

In relation to private sector TV production companies, which have become more and more important since the early 1990s, most are based in Beijing. China's first private TV production company, Jiashi Advertising (now renamed Jiashi Media) was set up in Beijing in 1994; it started with advertising production and now produces TV programs for CCTV and other TV stations. Nowadays, most of the large TV production companies are in Beijing. Having such a strong production force, in 2010 SARFT approved 122 shooting plans submitted by TV drama production organisations and companies in Beijing, accounting for 25 per cent of all TV dramas in China. Furthermore, in the film industry, nearly all the large private film companies are in Beijing, such as Huayi Brothers, Poly Bona, Stella Media Group (*xingmei chuanmei*), Beijing New Picture Film Co. (*xinhuamian yingye*), Orange Sky Entertainment Group (*chengtian yule*) and Enlight Pictures (*guangxian yingye*), in addition to two state-owned film companies, China Film Group and Beijing Forbidden City Film Co.

Finally, Beijing is also China's central hub of media business, including exports. The China Film Group and Huaxia Film Distribution Co. are the only two film companies certified by the government to import and distribution foreign films, and both are based in Beijing. The State Administration of Radio, Film & Television and China Media Group have jointly hosted the China Radio, Film & Television International Exposition in Beijing

TABLE 6.2
Major private TV drama production companies in Beijing

No.	Company name	Establishing Time
1	Hairun Movie & TV Group (*hairun yingshi*)	1993
2	Pigasus Movie Movie & TV Group (*jinyingma yingshi*)	1993
3	Yingshi Film & Drama Production Co.	1994
4	Huayi Brothers Media Group	1994
5	Rosat Film & TV Art Co. Ltd. (*rongxinda*)	1995
6	Beijing Beida Huayi Film & Drama Co.Ltd (Latterly merged into Media China Corporation Limited)	1996
7	Beijing Xinbaoyuan Movie & TV Investment Ltd Co.	1998
8	Beijing Galloping Horse Film & TV Production (*xiaoma benteng*)	1998
9	Ciwen Pictures	1999
10	Beijing Hualu Baina Film & TV Co. Ltd.	2002

Source: SARFT 2011

every year since 2003, which is now the largest film and TV trading and exhibition event in Asia. In 2011, the State Administration of Radio, Film and Television and the Beijing Municipal Government launched the Beijing International Film Festival as an international media showcase event, aiming to provide a platform for domestic and international film trade. The Ministry of Culture (MoC) has identified 59 key radio, film and TV export companies in China, and of these 18 are in Beijing; moreover, 43 of the 91 key national film, TV and radio export projects are in Beijing (MoC 2010).

THE CREATIVE MIGRATION TRAJECTORY IN BEIJING

As Curtin observes, during the pre-modern era in liberal democratic societies, creative migration was driven by patronage relations rather than market forces (Curtin 2007: 14). After the 1949 revolution, China's political centre shifted to Beijing, which represented the 'new China' spirit that appealed to many Chinese artists and writers, and who moved to Beijing in order to be more closely involved in the new Chinese culture of socialism. Beijing gradually re-established its status as China's cultural centre. As the economic reforms from 1978 onwards changed the social structure, Chinese people were increasingly able to move freely from job to job, and from place to place, and the role of the government employment assignment system receded. Over the past decade, creative people have been migrating to Beijing for better media-related education, more career opportunities, and to take advantage of Beijing's capital status, which has been established in their mind when they were children.

One factor driving creative migration to Beijing is the quality of its higher education institutions. Following decades of development, the Beijing Film Academy (BFA), the Central Academy of Drama (CAD) and the Communication University of China (CUC) have become the primary training grounds for most media professionals in China. As a result, many aspirants come to Beijing and attend the entry interview for these universities because they believe that entry into these universities is a precondition to have a successful career in the Chinese media industries. Indeed, many media practitioners described it as a simple and natural reaction to apply to BFA, CAD and CUC to pursue post-high school education. Employers hire graduates from BFA, CAD and CUC out of their confidence in these institutions' educational quality. Directors will often go to these universities when organising production teams for new projects. Students in these universities are seen as being 'ready-made' for China's films and TV industries. Moreover, many directors

and producers are bonded to these universities by virtue of their alumni relationships. As the bond connecting most Chinese media professionals, BFA, CAD and CUC are the most important entry points for aspirants seeking to enter into media careers. Li Yan, a freelance film director, quit school very early and tried several jobs, even playing rock music. Later on his interest shifted to media but he did not know how to start. He said:

> I didn't have any friend working in the media. No-one in my family ever worked in this area before. I didn't know how I can find a job in media industries and who I can go to. Then I thought maybe I should go to BFA or CAD to learn how to make films and during my study there I find the way to get myself in the industry as they have strong alumni networks. (Interview with author, 28 October, 2009)

A second factor driving creative migration to Beijing is the existence of strong local networks. By definition, the creative industries require highly flexible, open-ended and project-oriented production systems. Along with the accumulation of media in Beijing, more and more job opportunities are generated; this means there are more opportunities for subcontractors and freelance workers; and such employment is secured by being close to the largest pool of employment opportunities (Christopherson and Storper 1989). According to Wang Qiang, a native of Sichuan Province, 'People must come to Beijing because all the media companies, especially the big ones are in Beijing ... What every Chinese media worker really needs is a chance to realise his or her personal value... but you won't be able to work on one TV program or film if working in a small company' (interview with author, 29 October, 2009). Wang started his first company with friends while studying in CUC, although the company went bankrupt due to financial difficulties. He chose to stay in Beijing after graduation because of his networks and subsequently opened his second company with friends in TV, animation and advertising business. He believes that, by being in Beijing, he is exposed to more companies and organisations that he regards as potential clients, and his current client list includes CCTV, BTV, Heilongjiang TV, KAKU animation channel, CBN (the business channel of Dragon TV) and China Education TV.

Aside from the obvious business benefits, Beijing's status as the nation's capital lures Chinese people into its orbit. The capital culture of Beijing has an all-embracing spirit (*jianrong bingbao*). A TV producer at CCTV-2 said: 'I am from Jiangsu Province in the southern part of China. But I love the culture represented by Beijing which is so different from my hometown. It's so dignified and has grandeur' (interview with author, 12 November 2009). As the nation's power centre, Beijing symbolises the country's image, and this has a strong cultural influence on every Chinese, particularly as numerous cultural events are held in Beijing every year that get national exposure through CCTV. According to Ouyang Guozhong, an ex-producer at Hunan Satellite TV and CCTV, Chinese audiences are more interested in Beijing stories because of its dominance in Chinese culture (Ouyang 2003). Yang Li, a TV producer, was born in Xinjiang and had her college education in north-western Shaanxi Province. She recalled:

> I think I have a special complex for Beijing. When I was a little girl, I always wondered why I lived in Xinjiang instead of Beijing. At that time I watched CCTV all the time … Children in Beijing were able to perform at the opening ceremony of the Asian Games but I couldn't. They sang and danced on TV shows but I couldn't … Therefore Beijing to me is the centre of everything … Though I am not a Beijing native, I always feel so familiar with the city. (Interview with author, 4 November 2009)

THE MUTUAL LEARNING EFFECT IN BEIJING

The dynamics driving the media industry agglomeration in Beijing are in many respects government-driven. This is likely to be hard to sustain over the medium term, not least because the tendency of media reforms is towards increasing competition and the role played by market forces, which should in turn loosen the control of the state and ease the production environment, even if ideological and national media principles retain an overarching hegemonic status. But the mutual learning effect generated through continuous creative migration that is occurring in parallel with industrial accumulation gradually generates for Beijing what Allen Scott (2000: 166) refers to as 'favourable untraded interdependencies' or 'positive locational externalities'. By having a diverse range of creative talents living and working in close proximity, new knowledge is created while existing knowledge is transferred and spread among media workers in tacit and explicit ways. As a result, a locally based community of media workers evolves into an active hub of information exchange and knowledge production.

Institutions such as the BFA, CUC and CAD are one major location where knowledge transfers occur. Students study media knowledge in their coursework. But, more importantly, they learn tacit knowledge. The BFA, CAD and CUC have very strong relations with media industries. Many professors from these universities also work in the industry. Therefore their students can easily and quickly participate in the industry by means of internships or part-time work. Even after graduation, media talents still keep constant contact with their old schools to learn new knowledge.

However, studying at universities is still primarily associated with a one-way learning process. Most mutual learning effects actually happen during everyday work processes. According to Yang Li, a successful producer after working on several jobs in CCTV, BTV and private TV production companies, she absorbs new knowledge by working with different people from all kinds of backgrounds: 'Whenever I encounter with any difficulty I always talk to my old colleagues and learn from their experience' (interview with author, 4 November 2009). Nearly every new job for Yang Li was introduced by her old friends, who were her networks from previous work. She says that she learned a diverse range of production skills, concepts and ideas via collaborating with media workers from different organisations and institutions.

Likewise, Zhou Hao, the executive director of Golden Screen Television Media, explained that all his investors and partners were introduced by friends and he identified new ideas and opportunities through them. He said: 'We can't just work with someone in the market. We always collaborate with friends as we trust their capacity and moral qualities' (interview with author, 5 November 2009). He believes his social networks are very important to his work as they have various resources and knowledge. An expansive network brings in job opportunities as well as learning opportunities, according to Wang Qiang:

> We are good friends in life and competitors in work. I don't think they are contradictory to each other. At least, I am doing pretty well. We help each other by giving useful suggestions. Moreover there are plenty of job opportunities in Beijing and hence we share information rather than hiding from each other because in this big market, there are always something that they are good at and something that I am good at. More importantly, their work inspires me in two ways. On the one hand, I learn new technology by communicating with them and develop my own production format inspired by their work; on the other, their development gives me psychological pressure so that I always push myself to do better in order to catch up with them. (Interview with author, 29 October 2009)

The deepening of media reform has broadened channels for information flows, as more parties are brought into mutual learning processes. Yang Li described her old job in CCTV as a passive way of accepting her boss's instructions: she had no chance to express her thoughts to the boss (*lingdao*). But now, as a producer in a private TV production company, she is able to talk to CCTV when she produces programs for them. In this way, CCTV and she understand each other better and learn from each other. She says that state-owned TV stations realise that they have to adjust themselves in order to deal with the severe competition coming along with the reform. They need the help of private companies in order to be competitive (interview with author, 4 November 2009).

In these ways, an open-ended growth spiral is forming in Beijing, in which the centripetal logic of production is married to the centripetal trajectories of creative migration, as Curtin (2007) argues. In Beijing the university experience is not as dominated by pedagogy as might have been the case two decades ago. Institutions such as BFA, CAD and CUC are locales where informal study occurs and information intersects. Outside of the university, media workers spread knowledge spontaneously and consciously while they work and live collectively; accordingly new job opportunities emerge. The quickly developing industry and competition are catalysts accelerating the speed and scale of creative migration. At the same time, this creative migration drives economic expansion and industry agglomeration, which in turn enhances the place's attraction to a new generation of media workers.

THE CBD MEDIA CLUSTER AS THE SYMBOL OF BEIJING'S MEDIA CAPITAL ASPIRATION

In 2008, the Beijing Municipal government formally recognised the Beijing CBD International Media Industry Cluster, located around the new CCTV at the heart of the CBD in the eastern part of Beijing. The change to the media landscape of Beijing started with the announcement of CCTV's relocation plans. CCTV is the biggest TV station in China, and is considered one of the 'big three' media outlets in China, along with the *People's Daily* and *Xinhua News Agency*. CCTV has a close relation with China's State Administration of Radio, Film and Television (SARFT); it is the only TV station directly under SARFT, and normally the director of CCTV plays an important role in SARFT. As a result, CCTV always speaks for the central government, and its moves reflect the broader policy thinking of SARFT. Endowed with this special relation with SARFT, CCTV has been located in the west of Beijing, in Xicheng District, which is the locality of most central government branches. In this sense, CCTV's proximity to the government symbolised the political power behind China's media industries.

The CCTV centre in Xicheng had become too small to serve the station's future development plans. Many programs and people have to rent outside studios and offices, and the strict security measures created inconvenience for people entering the building, not to mention difficulties of coordinating such a huge but scattered institution. As a result, the relocation was inevitable for CCTV in order to continue its growth. In 2004, CCTV conceived a plan for moving to the core area of CBD. This decision arguably represents a strategic change in the nature of CCTV itself, from being primarily a party organ towards becoming an international media enterprise.

With the increasing commercialisation of the TV sector in China, and growing competition from domestic and foreign opponents, CCTV realised the urgency and necessity of reshaping its image and remaking its development strategy. Historically, as summed up by the old Beijing adage, the dwelling space of Beijing has been distributed in a way that 'in the east are the wealthy, in the west live the aristocrats and bureaucrats, while in the south and north are only poor people in the lowest level of the society' (*dongfu xigui nanpin beijian*). This spatial pattern is still inherited in Beijing today. The west urban area of the city, Xicheng District, is the site of most central government offices and military institutions, whereas the east urban area of the city, Chaoyang District, is the residence and working place for China's 'new rich class'. Hence, by moving to the CBD in Chaoyang District, CCTV was flagging its intentions to engage in further internationalisation and commercialisation.

Moreover, the unusual shape of the CCTV new building was also a striking visual representation of CCTV's ambitions of rising to be a world-class media organisation. CCTV is China's only national television network and is eager to build up a brand new image that can impact on the world, in accord with a national campaign of 'soft power' in which CCTV is a major force. The Dutch architect Rem Koolhaas's design was chosen by CCTV, and the construction of the new building began shortly after. The new CCTV building is a 234-metre high, 44-storey skyscraper in the Beijing Central Business District (CBD), whose construction cost 10 billion RMB, and it subverts the shape of CCTV's old building by being open rather than closed, and challenges conventional architectural ideas. The CCTV website describes the building as 'not only representing the image of a new Beijing, but also expressing the significance and cultural identity of TV media'.

But most importantly, the relocation of CCTV gave rise to a large-scale change to the media landscape of Beijing. Production companies, broadcasting institutions, advertising companies, training organisations and performance agencies which used to be in the west of Beijing began moving to the CBD soon after CCTV's relocation plan was released. A business relation with CCTV is widely seen as a symbol of strength; media companies collaborating with CCTV can benefit economically and from reputational effects. Beijing TV (the local TV station of Beijing) and Phoenix TV (the most popular overseas TV station based in Hong Kong) soon moved to the CBD. Despite the high rents, media companies have also moved to the CBD. Hence CCTV's relocation drove the restructuring of the landscape of media industries in Beijing, resulting in the emergence of the CBD International Media Cluster.

As an officially recognised creative cluster, the CBD International Media Cluster is strongly supported by the policies of the Beijing municipal government. In addition to the city's policies for creative industries and clusters, Beijing Municipal government also pays a lot attention to the development of media industries. In Beijing's Eleventh Five-year Plan (2006–2010), the Beijing government put forward the proposition of developing Beijing into the national centre of film and TV production and trade, and that the major measure for achieving this objective was to reform the system on a large scale. In this plan, the Beijing Municipal government designated Chaoyang District, especially the CBD area, as the key area for development of the media industries. It is written in the plan that 'based on CCTV and BTV, the CBD area will build a cultural and creative centre for media production, where cultural and media industries cluster'. The Chaoyang District government proposed in its Twelfth Five-year plan that the CBD will 'vigorously develop information and media industries … absorb more provincial and city TV stations … and enhance the cluster of international news agencies and media groups'.

The development of the CBD media cluster reflects the development process of Beijing becoming a media capital. In the first stage, the government initiates commercialisation reforms of the media industry and continues to deepen it, compelled by the developing demand and the pressure of competition. Various resources are redeployed via policies and administrative means to support state-owned media companies and organisations, especially the largest ones, who directly represent the government's interest in the media industry. Then private companies and foreign companies pursuing market opportunities seek to collaborate with state-owned media companies, which are normally under the government's protection, in order to reduce the risk engendered by the structured uncertainty of media policies. Media workers migrate continuously towards the new cluster and form a strong local network, while job opportunities increase along with the growth of the media industry overall.

BEIJING'S SOCIO-CULTURAL VARIATION

The final set of factors that Curtin raises in relation to the development trajectory of a media capital is that of what he terms a locality's forces of socio-cultural variation. In the case of Beijing, there are elements that support movement towards media capital status, and forces that retard such development. Most notable in the latter case are the contradictions that arise from Beijing's status as a political and administrative capital. While this serves to facilitate the movement of people and resources towards the city, as discussed above, it also frequently acts as a brake on innovation and creative thinking.

Beijing's dualistic intellectual culture

As China's national capital, Beijing has political, cultural and economic power. It has been the capital since the Yuan Dynasty in the thirteenth century, although China's political centre was moved to Nanjing in Jiangsu Province during the Kuomintang's brief regime in the twentieth century. Now Beijing is home to national regulatory agencies impacting on culture and audio-visual sectors such as SARFT and the Ministry of Culture. Therefore, for many outside China, Beijing is seen as being symbolic of China's conservative political ideology: its opulent palaces and temples, huge stone walls and gates present an image of stability over time.

Paradoxically, Beijing is also the hotbed of China's modern culture. It is symbolic of a rebellious spirit. Historically, Beijing is the origin of the New Culture Movement (*xin wenhua yundong*) and the May Fourth Movement (*wusi yundong*) in mid 1910s and early 1920s, which led a revolt against Confucian culture, calling for the creation of a new Chinese culture based on global and western standards, especially democracy and science. In modern times, Beijing's cultural formations have shown resistance to power. Beijing's so-called rural-urban syndicates *(chengxiang jiehebu)* are formations of contemporary artists who represent the 'cynical realism' movement, a school that mocks China's political and social realities, while in Beijing's cafés and bars one finds independent rock bands expressing their angst about government and social conditions. In the late 1980s and early 1990s, Beijing was the centre of the 'hooligan' literature movement, best represented by Wang Shuo, a writer who symbolised Beijing people's suspicion of authority.

Beijing's intellectual culture includes a significant part of its so-called elites. As a privileged stratum, they have long been exposed to the world via media and by reading literature not accessible to common people. With personal experience of the political system and material wealth inherited from their families, they have had the luxury of making culture, a situation reminiscent of the period after 1911 which led to the New Culture Movement, when court intellectuals were released from their responsibility. For instance, Wang Shuo grew up in a military family in Beijing; his work was made into several successful TV dramas and films in China. According to Tom Wang, CEO of Beidong Media, while Shanghai or Guangzhou may have better performance in economic development, 'Beijing has inherited an atmosphere of culture ever since the Ming Dynasty and people living in Beijing want to speak out instead of earning money' (interview with author, 23 October 2009). Xu Gang, deputy director of distribution of August First Film Studio (*bayi dianying zhipian chang*) commented: 'It's so different in Beijing.

Here even a taxi driver criticises current political affairs and makes jokes of the central leaders. But in Shanghai, people are busy making money and have no time to discuss or even think about other things' (interview with author, 19 November 2009).

The political capital status

As many scholars have noted (Chan 2003; Fung 2009; Sun and Zhao 2009), China's media has been granted a commercial mandate while fulfilling their function as party mouthpiece. Inevitably the government is an important factor influencing the growth and clustering of media industries across the country. Ideology remains a significant consideration, not only for media companies' survival, but also for government officials in deciding whether or not a TV program or film can be broadcast. China has thus been described as a 'treacherous terrain' (Curtin 2010a: 24) for media companies, with ambiguous policies, political favouritism, censorship, piracy and a lack of financial transparency among the most common concerns. As the political centre of China, Beijing provides the best access to critical information for lowering business risks; in China this critical knowledge includes government policy developments. Understanding cultures of production in Beijing is important for both local and international players in reducing transaction costs.

At the same time, an 'adaptive informal institution' (Tsai 2006) operates as a result of the close connection of the government and media companies in Beijing which facilitates the institutionalisation of those informal coping strategies to formal institutions through processes of repetition and diffusion, and even enables political elites to reform the original formal institutions. The adaptive informal institutions are embodied in various acts based on social relationships (*guanxi*) in the Chinese society. In China, the reform process has led to a significant expansion of *guanxi,* rather than its decline in place of more contractual modes of interaction. Though the *guanxi* between entrepreneurs and government officials is responsible for the corruption often seen in China, it remains an indispensable facilitator for entrepreneurs to secure their business interests in the insecure social and political environment in China. Accordingly, many media enterprises in China need to make constant contacts with the government agencies such as SARFT in order to push their own interests. *Guanxi* not only can provide advance notice about policy changes, but can also enable prior adaptation to such changes.

Furthermore, China's bureaucratic system does not include a formal channel through which people can voice their opinions on current or proposed changes, and hence be more directly involved in the decision-making process.

In Western countries, industry associations have developed the capacity of lobbying the government in order to influence public policies. But there is no presence of strong industry associations or groups in China, since all are established and managed by the Communist Party of China (CPC). Therefore, the creation and use of *guanxi* is commonly adopted by entrepreneurs as a way of influencing policy-makers' decisions, to negotiate with government, and sometimes even persuade government, in order to push reform in the direction they want. Especially in China's media regulatory system, there is no particular law or Act for media industries. Rather, SARFT constantly changes media-related policies and regulations in order to maintain a harmonious society and, more importantly, to serve the central leadership's ideas and objectives. But instead of a routinised and professional bureaucratic system as has come to emerge in Western countries, policy-making and implementation in China remains a highly specific and personalised endeavour (Breznitz and Murphree 2011). The lower the official's level in the bureaucratic system, the more conservative she or he is. In order to keep their positions in the government, officials have to play safe and interpret central leaders' intentions cautiously to prevent any problems which may irritate central leaders; officials on the execution level do not dare to test the boundary of the central leaders who are several levels higher than them. In addition, the different interest groups existing within the Party increase the complexity of policy-making and implementation.

Under these circumstances of 'ruling by person', the question of media policies is in fact an issue of politics. One can win the first-mover advantage and take the lead by exerting a bigger influence on government via adaptive informal institutions such as *guanxi*. Over time, those isolated adaptive responses of the government to the industry development evolve into the adaptive informal institutions, or what Tsai describes as the 'regularized patterns of interaction that emerge as adaptive responses to the constraints and opportunities of formal institutions, that violate or transcend the scope of formal institutions, and that are widely practiced' (Tsai 2006: 125). In this sense, adaptive informal institutions bridge the two ends together – media companies that need policy support from the central government, and the central government, especially the central leaders, who are eager to develop China's media industries. Hence, media professionals need to be in Beijing to establish *guanxi* with government officials in order to negotiate for more interests. More importantly, they benefit from being in Beijing because during interactions with government officials the adaptive informal institutions gradually form and 'your feedback is appreciated by the government and impacts on their future measurement … in this way your desired changes will happen sooner or later' according to Qian Chongyuan, the distribution director of Beijing Forbidden City Film Company (interview with author, 26 November 2009).

Facilitated by Beijing's dualistic intellectual culture and its political capital status, a stable path of industry accumulation can be formed. Curtin argues that the dynamics of media agglomeration are stable once they are formed, and that locales that fail to make an early start are thereby subject to 'lock-out' (Curtin 2010a: 5). Therefore, the agglomeration of media industries in Beijing would seem to be likely to continue, since the choice of coming to Beijing is made by new firms who are heavily influenced by the previous development in Beijing as a result of prior path-dependence, similar to the formation of Silicon Valley as described by Arthur (1990).

CONSTRAINED CAPACITY TO INNOVATE AND EXPORT

While it seems that Beijing is the destination which Chinese media resources and talents are flowing to, for the reasons outlined above, the close connection with politics clearly constrains the Chinese media industry's capacity to innovate and export. By creating a complicated operating environment which helps to keep foreign capitalists at bay, the Chinese government intends to gain time for its domestic media industries to grow strong enough to fight with Hollywood media conglomerates. But such a fickle environment discourages media talents who require a more stable policy environment, and undermines the development of media enterprises, both domestic and international.

Furthermore, Curtin argues that media capitals emerge where opportunity, prosperity, expressive freedom and rich cultural resources converge. He argues that a city's ability to attract creative labour and make effective use of it seems to rely on not the city's supply of creative venues but also their diversity. Similar points are made by John Montgomery (2007; see also this volume). But in China, an over-powerful government and its extensive policy leverage leads to a lack of space for diversity of content and expression. Consequently, China's media industry is not able to produce content that is sufficiently appealing to audiences in markets other than that of mainland China. Zhang Yimou and his peers in Beijing therefore remain frustrated that their work is not received as well in the international market as they believe that it should. In the long term, the government may not be able to distribute resources to the places where it is mostly needed despite its strong capacity to concentrate capital and talent.

Though Beijing is a city where we see a great deal of cultural diversity, the key problem is how much cultural diversity is allowed by the Chinese government to be incorporated into media content, which remains the government's primary ideological control tool. Looking at media policies and the production environment in China, it is evident that the answer to the question remains 'not much'. The foremost regulatory measure adopted by the Chinese government is censorship. Compared with international media environments, a TV drama or film needs to pass through censorship procedure twice in China. One is before production – referred to as 'registering in the system' (*beian gongshi*) in relevant media policies – and the other is after production. According to China's media regulations, a production company needs to submit production details including storyline and plot to SARFT for registration prior to the production. The production organisation will be asked to stop production if SARFT finds something wrong or in conflict to its expectations concerning culture in the production plan. In this regulation system of double censorship, topics are filtered twice by the government – both prior to production and after production – and only topics that are in line with the culture the government *wants to promote* can pass through the SARFT process.

In terms of media exports, the Chinese government controls what content can be exported to overseas audiences. The Chinese government commonly wants to show the globe the prominent traditional Chinese culture and the happy life that Chinese people enjoy in modern times. Media export is strongly associated with the soft power strategy. Therefore it is impossible for the Chinese government to accept any content that does not conform to the mainstream ideology in exported media product, not to mention mocking or criticising the state.

With this constrained capacity to innovate and export, China's media industries are still develop in an inward-looking way: the government still implements its own rationales in the regulation practice due to the ideological concerns, and most media companies still focus on the domestic market although they claim that they have made plans for internationalisation. The production quality is frequently compromised by the pursuit of short-term interests. Yin Hong has observed that only 2/5 of the films produced can be exhibited in cinemas and circulated in the market every year, although the production yield grows quickly (Yin and Cheng 2011). In the TV sector, Hu Zhengrong and his colleagues have also found that re-broadcasting old programs is rampant in China's TV industry, that China's TV sector is full of program cloning and homogeneity, and that it severely lacks innovation and creativity (Hu *et al.* 2011).

This chapter has sought to apply Michael Curtin's concept of media capitals in order to consider the relevance of its three core concepts – logics of economic accumulation, trajectories of creative migration and forces of socio-cultural variation – can be applied in the case of

Beijing. It found that the first two of these forces are strong in Beijing, and that state-sponsored industrial cluster formation in the CBD area of Beijing is shaping growth in the media sectors in the city, developing a dynamism that has the potential to be self-sustaining, as has been the case with archetypal media capitals such as Hollywood.

At the same time, the extent to which this is government-driven, combined with the dual nature of China's media industries as commercial entities and propaganda instruments, generates recurring difficulties in building the momentum required to establish Beijing as a globally recognised media capital, rather than the primary site of Chinese film and television production. The nature of reform in China, in the media as in other sectors, has been described as a 'trial-and-error [process of] economic experimentation led by subnational entities but fashioned by political contestations between conservatives and reformers at the centre' (Breznitz and Murphree 2011: 20). Processes of informal institutional adaptation and the continuing centrality of *guanxi* relationships mean that path-dependent progress and reform continues to occur, but the balance between a bureaucratic-authoritarian state that sees itself as having a particular guiding role towards the media, and the imperatives of media marketisation and the aspirations to establish Chinese media companies as global players, continue to generate tensions and contradictions.

For commercial media companies, the economic opportunities presented by the size and growth of the Chinese media market remains tempting, whatever the practical difficulties of doing business. Under the current circumstances, media companies must be sensitive to official policies, which invariably aim to insure state leadership and the supremacy of the Communist Party in order to succeed in the mainland market. The difficulty is that this fealty to the party-state continues to undercut the exportability of local productions. Furthermore, Beijing's formal and informal political institutions and spheres of uncertainty shape growth in ways that are oriented towards short-term interests, and do not particularly reward creativity and risk-taking. But, although still influential, the territorial power of government policies over media circulation and consumption decreases with the continued process of marketisation.

Is Beijing a global media capital in making? The answer is yes in terms of scale. Defended by the government's policy leverage and fashioned with Beijing's own cultural specificity, media companies and talent will continue to aggregate in Beijing in the foreseeable future. The problem, however, is that even with the joint efforts of the government and the media industry, Beijing has not been able to achieve the kind of success that the authorities desire, and is not yet a media capital as its media exports remain small, and the cultural influence of its media on other countries or regions remains limited. Therefore, instead of protectionism, the government should focus on supply, encouraging the capacity to promote and sustain creativity and innovation in its media sectors, if the government wishes to continue to pursue the clustering strategy.

REFERENCES

Arthur, W. B. (1990). 'Silicon Valley' locational clusters: when do increasing returns imply monopoly? *Mathematical Social Sciences*, 19(3): 235–51.

Breznitz, D. and M. Murphree. 2011. *Run of the red queen: Government, innovation, globalization, and economic growth in China*. New Haven: Yale University Press.

Chan, J. M. 1993. Commercialization without independence: Trends and tensions of media development in China. In *China review 1993*, eds M. Brosseau and J. Y.-S. Cheng. Hong Kong: Chinese University Press.

Chan, J. M. 2003. Administrative boundaries and media marketization: A comparative analysis of the newspaper, TV and Internet markets in China. In *Chinese media, global contexts*, ed. C. C. Lee. London: RoutledgeCurzon.

Chen, G. 2004. *Bohemia China*. Guilin: Guangxi Normal University Press.

Christopherson, S. and M. Storper. 1989. The effects of flexible specialization on industrial politics and the labor market: The motion picture industry. *Industrial and Labor Relations Review* 42(3): 331–47.

Curtin, M. 2003. Media capital: Towards the study of spatial flows. *International Journal of Cultural Studies*, 6(2): 202,en>28.

Curtin, M. 2007. *Playing to the world's biggest audience: The globalization of Chinese film and TV*. London: University of California Press.

Curtin, M. 2010a. *Chinese media capitals in global context*. Paper presented at the 60th annual meeting of the International Communication Association, Suntec Singapore International Convention & Exhibition Centre, Suntec City, Singapore, 21–25 June.

Curtin, M. 2010b. Comparing media capitals: Hong Kong and Mumbai. *Global Media and Communication*, 6(3): 263–70.

Flew, T. 2007. *Understanding global media*. Basingstoke: Palgrave Macmillan.

Fung, A. Y. H. 2009. Globalizing televised culture: The case of China. In *Television studies after TV: Understanding television in the post-broadcast era*, eds G. Turner and J. Tay, 178–88. London: Routledge.

Hu, Z., J. Li and W. Huang. 2011. The development of radio and TV industry in 2010. In *Report on Development of China's media industry*, ed. B. Cui, 147–56. Beijing: Social Sciences Academic Press (China).

Hui, D. 2003. *Baseline study on Hong Kong's creative industries*. Hong Kong: Centre for Cultural Policy Research, The University of Hong Kong.

Lee, C.-C. 2009. *Chinese media, global contexts*. London: Routledge.

Liu, K. 2009. Creative edge of cities: A comparative analysis of the top 500 creative industries businesses in Beijing and Shanghai. *Creative Industries Journal*, 1(3): 227–44.

Miller, T, N. Govil, J. McMurria and R. Maxwell. 2005. *Global Hollywood 2*. London: BFI.

Ministry of Culture (MoC). 2009. *Catalogue of cultural export enterprises of excellence in China (2009-2010)*. http://fwmys. mofcom.gov.cn/aarticle/a/ad/200911/20091106638174. html?1741152766=201099483.

Ministry of Culture (MoC). 2010. *Catalogue of cultural export projects of excellence in China (2009-2010)*. http://www.mofcom.gov.cn/ aarticle/bh/201001/20100106740265.html? 1137107454=201099483 ().

Montgomery, J. 2007. *The new wealth of cities: City dynamics and the fifth wave*. Aldershot: Ashgate.

Ouyang, G. 2003. *The great transition of Chinese media*. Beijing: Tuanjie Press.

Scott, A. J. 2000. *The Cultural Economy of Cities*. London: Sage.

Sina.com (2011). *2010 China film top 10*. http://ent.sina.com.cn/m/ c/2011-01-13/16303206711.shtml (accessed 2 May, 2011).

State Administration of Radio, Film and Television (SARFT). 2011. *The development of radio, film and TV in 2010, and the plan for 2011*. http://www.sarft.gov.cn/articles/2011/01/13/201101131657 30840340.html (accessed 2 May, 2011).

Storper, M. 1997. *The regional world*. New York: Guildford.

Sun, W. and Zhao, Y. (2009). Television culture with 'Chinese characteristics': The politics of compassion and education. In *Television studies after TV: Understanding television in the post-broadcast era*, eds G. Turner and J. Tay, 96–104. London: Routledge.

Tan, Y. and Z. Wang, 2009. An analysis on economic and geographical differences of Chinese media economy. *Economic Geography* 29(3), 446–49.

Tsai, K. S. 2006. Adaptive informal institutions and endogenous institutional change in China. *World Politics* 59(1): 116–41.

Yin, H. and W. Cheng. 2011. Chinese film industry in 2010. In *Report on development of China's media industry (2011)*, ed. B. Cui, 183–200. Beijing: Social Sciences Academic Press.

Yu, G. (ed.) (2010). *Annual report on China's media development index (2010)*. Beijing: People's Daily Press.

Zhang, Y. (2009). *Go north or die*. http://www.infzm.com/ content/27329 (accessed 4 May, 2011).

Creative Industries after the First Decade of Debate

Terry Flew and Stuart Cunningham

Creative Industries Faculty, Queensland University of Technology, Brisbane, Queensland, Australia

It has now been over a decade since the concept of creative industries was first put into the public domain by the Blair Labour government's Creative Industries Mapping Documents in Britain. The concept has gained traction globally, but it has also been understood and developed in different ways in Europe, Asia, Australia, New Zealand, and North America, as well as through international bodies such as UNCTAD and UNESCO. A review of the policy literature reveals that although questions and issues remain around definitional coherence, there is some degree of consensus emerging about the size, scope, and significance of the sectors in question in both advanced and developing economies. At the same time, debate about the concept remains highly animated in media, communication, and cultural studies, with its critics dismissing the concept outright as a harbinger of neoliberal ideology in the cultural sphere. This article couches such critiques in light of recent debates surrounding the intellectual coherence of the concept of neoliberalism, arguing that this term itself possesses problems when taken outside of the Anglo-American context in which it originated. It is argued that issues surrounding the nature of participatory media culture, the relationship between cultural production and economic innovation, and the future role of public cultural institutions can be developed from within a creative industries framework and that writing off such arguments as a priori ideological and flawed does little to advance debates about twentieth-century information and media culture.

INTRODUCTION: CREATIVE INDUSTRIES POLICIES AROUND THE WORLD

The concept of creative industries emerged in the late 1990s primarily as a policy discourse, although the subsequent decade has seen a lively set of academic as well as industry and policy-related debates about its utility and implications for research, criticism, and creative practice. Its origins can be traced to the decision of the then newly elected British Labour government of Tony Blair to establish a Creative Industries Task Force (CITF), as a central activity of its new Department of Culture, Media and Sport (DCMS). The story of the CITF has been told by various sources (Hartley 2005; Pratt 2005; O'Connor 2007; Hesmondhalgh 2007) and is not recounted in detail here, except to note its four major contributions.

First, it established the creative industries as a central plank of the United Kingdom's "postindustrial" economy, observing that the sector accounted for 5 percent of total national income in 1998, employed 1.4 million people, and was growing at about double the rate of the British economy as a whole. Estimates in the United States were that the creative industries account for 7–9 percent of gross national product (Americans for the Arts 2008; Siwek 2006), while countries as diverse as Australia, Singapore, South Africa, and China were identifying figures in the range of 3–5 percent (UNCTAD 2008; Cunningham and Higgs 2008). Second, it marked out the continuation of a trend, first identifiable in cultural policy in the United Kingdom and Australia in the 1980s and early 1990s, to view cultural sectors as contributors to wealth creation and economic performance, and not simply as claimants on public revenues on the basis of nonmarket or intrinsic values. Third, by approaching the creative industries in ways that went beyond the traditional discourses of the subsidized arts, and giving a central role to creativity in the generation of economic wealth, debates about these sectors moved into larger discourses such as those of trade policy, copyright and intellectual property, urban development, and educational futures. Finally, in developing a diverse and eclectic list of industries that ranged from commercial media to publicly subsidized arts, the live-analog and the digital-multimedia, and those with largely one-off artisanal modes of cultural production to complex and highly capitalized sites of cultural production, creative industries were explicitly linked to discourses surrounding technological convergence, the information society and the "new economy" (Flew 2005a).

The creative industries policy discourse was taken up in a number of other countries. Singapore and Hong Kong developed detailed analyses of their creative industries sectors that were strongly influenced by the UK model. The concept also gained policy purchase in Taiwan, Korea, and, in the hybrid form of "cultural creative industries," in China (Kong, Gibson, Khoo, and Semple 2006; Keane 2007). The European Union (EU) identified the cultural sectors as experiencing employment growth that was four to five times the EU average, and noted that the cultural workforce was pioneering wider trends in European labor markets, such as higher rates of self-employment, high levels of tertiary education, and a greater proportion of the workforce in temporary or contract jobs (KEA 2006). Most European governments were hesitant to adopt the British formulation of creative industries, preferring to talk of the cultural industries or the cultural sectors, whereas some Scandinavian countries talked of the creative economy or the experience economy. Creative industries was also taken up by governments in Australia and New Zealand, although in the Australian case it was state governments, such as the Queensland government, that were the more enthusiastic proponents of creative industries policies (Craik 2007). In the United States, the comparative weakness of national cultural policies is offset by a patchwork of subnational strategies, where state and local governments undertake diverse initiatives to bolster the arts and entertainment industries, often with the purpose of rebadging their city or region as a hub of creativity (Wyszomirski 2008). On a global scale the United Nations Commission on Trade, Aid, and Development (UNCTAD) has become an enthusiastic proponent of the creative industries as a new engine of growth in developing countries (Barrowclough and Kozul-Wright 2008; UNCTAD 2008), while the United Nations Educational, Scientific, and Cultural Organization (UNESCO) has significantly upgraded its statistical frameworks to incorporate the size, scope and significance of cultural production in the global economy (UNESCO 2007).

THE STRUGGLE FOR DEFINITIONAL COHERENCE

The early list-based approach to creative industries developed in the UK context by DCMS was open to the charge of ad hocery, as it was not clear what were the underlying threads linking this seemingly heterogeneous set of industry subsectors (Flew 2002). The list drew together industries that were highly capitalized and industrialized in their modes of production and distribution (e.g., film and television), and those that were more labor-intensive and artisanal (arts and crafts, designer fashion, music, the visual and performing arts), as well as combining highly commercial sectors strongly affected by the business cycle (e.g., advertising, architecture), with arts sectors largely driven by public subsidy. Critics of the DCMS approach, such as Nicholas Garnham (2005), argued that the inclusion of the software, computer games, and electronic publishing industries had the effect of artificially inflating the size and economic significance of the creative industries, while David Hesmondhalgh (2007) questioned the exclusion of sectors such as heritage, tourism, entertainment, and sport.

The two lines of demarcation proposed in the original DCMS definition—individual creativity that could take the form of intellectual property—did little to clarify what Andy Pratt (2005) described as the "breadth question" in differentiating the creative industries from other sectors, while Chirs Bilton and Ruth Leary (2002, 50) observed that "it is difficult to think of a product which does not exploit some intellectual component in the form of patents, design elements or other intangible, symbolic properties which make that product unique," The focus on intellectual property also raised concerns that in the shift from cultural to creative industries, the result would be a one-sided focus upon that which was "new," produced through digital technologies, and commercially oriented, losing sight of the complex cultural ecologies that link commercial and publicly supported form of cultural production, as well as links between digital and tangible arts and media forms. This focus upon the link between creativity and intellectual property also raised concerns about the risks of subordinating culture to the commercial market in the shift from cultural to creative industries (Hesmondhalgh 2008).

The 2000s have seen increasing consensus emerge among policymakers about working definitions of the creative industries and what sectors should or should not be included, albeit with some debates about what should be considered to be "core" creative or cultural industries. Work undertaken for UNESCO in developing its revised Framework for Cultural Statistics (UNESCO 2007) has compared the cultural statistics frameworks of fourteen constituencies (ten countries, one region, one city, the European Union, and the World Intellectual Property Organization) and has found that a consensus exists around inclusion of the following sectors in cultural statistics modeling:

1. Publishing and literature.
2. Performing arts.
3. Music.
4. Film, video, and photography.
5. Broadcasting (television and radio).

6. Visual arts and crafts.
7. Advertising.
8. Design, including fashion.
9. Museums, galleries, and libraries.
10. Interactive media (Web, games, mobile, etc.).

Sectors for which there continues to be debate about their inclusion in a cultural statistics framework include architecture, software, product and reception hardware (e.g., musical instruments, electronic goods), festivals, intangible cultural heritage, and leisure activities, including sport. Nonetheless, considerable progress has been made in developing common statistical and analytical frameworks that enable the development of empirically robust data that can be compared between countries and analyzed over time, thus allowing for more speculative claims to be subject to detailed policy-related performance metrics. This can be seen in the work of UNCTAD on creative industries and the creative economy, which has been able to develop a sectoral taxonomy of the creative industries between the arts, media, heritage, and "functional creations" (or more service-oriented sectors), around the following broad definition of the creative industries:

- The cycles of creation, production, and distribution of goods and services that use creativity and intellectual capital as primary inputs;

- A set of knowledge-based activities, focused on but not limited to the arts, potentially generating revenues from trade and intellectual property rights;

- Tangible products and intangible intellectual or artistic services with creative content, economic value, and market objectives;

- At the cross-roads among the artisan, services, and industrial sectors; and

- Comprising a new dynamic sector in world trade. (UNCTAD 2008, 13)

Figure 1 shows how these principles link up to sectors and categories that constitute the creative industries worldwide.

The question of what relationship exists between the public and private sectors in creative industries development has also been clarified significantly. Rather than being a discourse that simply champions commercial popular culture as the obverse of traditional "market failure" rationales for arts and cultural funding, what has instead been emerging is a better understanding of the enabling role of public-sector institutions and government-funded cultural activities as drivers of innovation and socially networked markets (Cunningham et al. 2008). John Holden has observed that "as greater numbers of people are engaging with the content and spaces of publicly-funded culture . . . the working lives of greater numbers of people are taking on the characteristics and processes of cultural practitioners" (Holden 2007, 8). The public sector has typically been seen as both an enabler of commercial creative industries, assisting with the development and provision of inputs, including people with creative skills, as well as

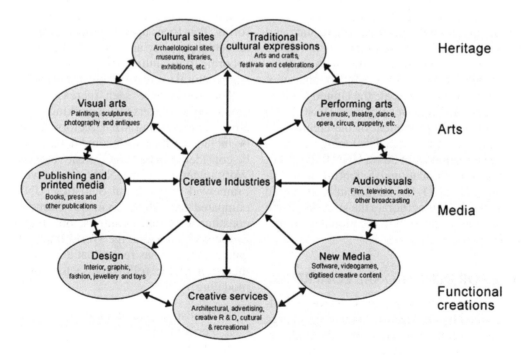

FIG. 1. United Nations Committee on Trade, Aid, and Development model of the creative industries. Source: United Nations Committee on Trade, Aid, and Development, *Creative Economy Report 2008* (Geneva: Author, 2008).

having responsibility for the empirical "mapping" of these sectors and their economic impacts (Holden 2007). Since the global financial crisis of 2008 and the subsequent economic downturn, public-sector organizations are increasingly being looked to for taking the lead on innovation in the creative industries sectors (Pratt 2009).

CULTURAL AND CREATIVE INDUSTRIES: INFORMATION SOCIETY REDUX?

Cultural economist David Throsby (2008) has observed that the distinctions made between cultural and creative industries arise from a mix of inherent definitional difficulties in delineating "culture" and "creativity"; distinctive national traditions in understanding these categories; and the politics of cultural policy and the impact of government priorities at the time on how the relevant sectors are defined and their roles conceptualized. Such questions arise, for Throsby, in a context where "the very status of cultural policy has been changing in a number of countries as a result of the emergence of the cultural industries as an object of interest to economic policy-makers"; a growing interest in culture as a source of economic value-adding is seeing "cultural policy . . . rescued from its primordial past and catapulted to the forefront of the modern forward-looking policy agenda, an essential component in any respectable economic policy-maker"'s development strategy" (Throsby 2008, 229).

Cultural economists often use the terms *cultural* and *creative industries* interchangeably, as their primary focus is upon the mix of product qualities and consumer needs that renders a particular commodity, activity, or service to be deemed cultural (Towse 2003). In other accounts, most notably the work of Throsby (2001, 2008), arts and cultural industries are the respective core subsets of the creative industries in what is known as the "concentric circles" approach, where industries are distinguished by the "core" role given to creativity in the input stage of production (e.g., the visual arts would be seen as a "core," but advertising would be seen as more "peripheral" as it combines creative inputs with other inputs). From the economic perspective, industry classifications are loose taxonomies that have to fit upon the shifting nexus of supply-demand relationships between individuals and firms; the classification of commodities and markets is analytically prior to that of industries.

In the critical humanities, by contrast, the political and ideological weight given to these different signifiers of an industry is substantially greater. The term "cultural industry" has its origins in neo-Marxist philosophy, where the rise of industrialized cultural production in the early twentieth century was seen—and is still seen by many critical theorists—as promoting "class rule," a "capitalist lifestyle" and an "administered society" (terms taken from Edgar 2008). There has, however, been a turn away from a monolithic and pessimistic model of "culture industry" since the 1970s, marked not only by a conceptual shift to a more empirically based understanding of how production, distribution, and circulation actually worked in these quite varied industries, but also by a greater interest in the policy settings that could enable these sectors to grow while furthering social-democratic agendas to democratize access to and participation in the cultural sphere. In his highly influential work for the left-wing Greater London Council in the early 1980s, political economist Nicholas Garnham argued against what he termed the "idealist" tradition in cultural policy that rejected markets as incompatible with culture, pointing out that "most people's cultural needs and aspirations are being, for better or worse, supplied by the market as goods and services" (Garnham 1987, 25). Garnham instead argued for approaches that would better understand how the cultural industries worked as "institutions . . . which employ the characteristic modes of production and organization of industrial corporations to produce and disseminate symbols in the form of cultural goods and services generally, although not exclusively, as commodities" (ibid.). The intention was to enable cultural policymakers to better identify the scope to intervene in cultural markets in order to further access and participation agendas, rather than reject cultural industries and the commodity form *tout court*.

It is worth noting, then, that Garnham subsequently came to be one of the more trenchant critics of creative industries theory and policy discourse, claiming that it inappropriately smuggles in arguments and assumptions associated with the information society. Garnham (2005, 20) agues that the shift from cultural to creative industries is opportunistic, "an attempt by the cultural sector and the cultural policy community to share in its relations with the government... the unquestioned prestige that now attach to the information society and to any policy that supposedly favors its development." Moreover, in the British context of Tony Blair and Gordon Brown's "New Labour," creative industries was perceived as a policy Trojan horse to promote "the shift to and reinforcement of 'economic' and 'managerial' language and patterns of thought within cultural and media policy" (ibid., 16). This critique of creative industries as a form of ideological mystification has also been made by David Hesmondhalgh (2008, 567), who proposed that it is based upon "arguments which all too often come close to endorsing inequality and exploitation associated with contemporary neo-liberalisms," and by Andrew Ross, who suspected the new policy rubric of being "'old wine in new bottles'—a glib production of spin-happy new Labourites, hot for naked marketization but mindful of the need for socially acceptable dress" (Ross 2007, 18).

The creative industries concept shares with information society theories an interest in the long-term shift in employment and national income from agriculture and manufacturing to services in advanced capitalist economies; the limitations of "industrial era" statistical modeling in capturing the economic dynamics of services and information-based sectors; and the growing role of knowledge capital as a primary driver of growth in these economies (David and Foray 2002). In terms of the five sets of arguments that Garnham associates with information society thinking (for another accounts of the information society, see Hassan 2008), theories of the service economy and post-Fordist flexible production models have been most influential, with transaction cost-based theories of the firm helping to shape some economic accounts of creative industries (e.g., Caves 2000), and Schumpeterian models of innovation and entrepreneurship influencing thinking about the significance of small-to-medium firms as incubators of innovation and new business models in the arts, media, and cultural sectors (Cunningham et al. 2008).

Notions of the Internet and digital media as "technologies of freedom" have had more influence on the growing literature on the significance of networks and collaborative social production (e.g., Benkler 2006) than on creative industries theories. Daniel Bell's original thinking about the information society has had only limited influence, with most accounts of the creative industries being considerably more circumspect about proclaiming a postindustrial utopia for the creative workforce, and there has been considerable recognition of the question of precarious labor that often underpins contract employment and flexible working arrangements (Deuze 2006; Cunningham 2008). Indeed, creative industries theorists have often pointed to the limitations of information society models, arguing that a focus on human creativity—often derided among critical theorists as a residue of bourgeois individualism and a romanticisation of the artist (Miller 2002; Donald 2004)— is an important corrective to policies overly focused upon infrastructure and technological hardware to the detriment of human factors and social software (Mitchell et al. 2003; Hartley 2005; Cunningham 2009b).

Garnham's (2005, 16) claim that creative industries discourse "assumes that we already know, and thus can take for granted, what the creative industries are, why they are important and thus merit supporting policy initiatives" is also open to question, particularly if we shift our focus outside of its origins in Great Britain under Tony Blair's New Labour. Cunningham's (2007; 2009b) evaluation of 1200 creative industries policy documents developed outside of the United Kingdom from 1998 to 2006 found a high degree of incrementalism, policy variance, and attention to local contextual factors, rather than an unreflexive blanket imposition of neoliberal rational–comprehensive orthodoxies about the proper culture–economy relationship. In very general terms, a global scan of this policy literature finds four main variants:

- A United States model, where there is a substantive divide in thinking and calculation toward arts and culture on the one hand and the entertainment/copyright industries on the other, and where the bulk of policy initiatives are highly localized and subnational in their focus, as seen with the rise of the "creative cities" movement (Wyszomirski 2008).

- A European model that emphasizes the cultural mission of these industries and strategies for social inclusion for common cultural benefit and where the term "cultural industries" is generally preferred to that of creative industries.

- A diverse range of Asian approaches, which strongly emphasize the role of national sociocultural and political circumstances, but still identify opportunities for export growth and successful branding of global city-region in the highly competitive Asia-Pacific region, while at the same time challenging long-held orthodoxies about instrumentalist education and the dominance of the ICT sectors in driving economic growth (cf. Kong et al. 2006).

- Developing country models in South America, South Africa, the Caribbean and elsewhere, where questions of cultural heritage maintenance, poverty alleviation, and provision of basic infrastructure have precluded overly technocratic conceptions of creative industries being promoted uncritically as the inevitable fruits of the information society (UNCTAD 2008).

CREATIVE INDUSTRIES AND CULTURAL POLICY

What has been traced in this article thus far is the manner in which the rise of creative industries as both theory and policy discourse has intersected with changing understandings of the relationship of the arts, the media and applied creativity to new media technologies, globalization and the twenty-first century knowledge-based economy. Over the period from the initial development of creative industries in the United Kingdom in the late 1990s to the present, there has been a refining of definitions, models, agreements of what industries are included or excluded, and measurement techniques. This has reached the point where recent work undertaken through bodies such as UNESCO (2007) and UNCTAD (2008) is generating more statistically robust data on the size, scope, and significance of the creative industries on a global scale, which rests upon stronger empirical foundation than the more ad hoc or speculative accounts that prevailed in the late 1990s. At the same time, there are significant variations in national and regional adaptations of the creative industries template from the form in which it first emerged in the United Kingdom, in contrast to claims that it is simply the reflection of a singular "master discourse," such as the information society, exported from Tony Blair's "Cool Britannia." A distinction that could be further explored is the difference in applications of the concept between Europe and Asia. Europe has tended toward what has been referred to as the "concentric circles" model, where arts-related activities are seen as being in the "core" creative industries, whereas fields such as advertising, architecture, and design as well as media industries are seen as only being partially creative (Throsby 2001; KEA 2006; Work Foundation 2007). By contrast, Asian definitions of the creative industries have tended to be more eclectic and inclusive. Indeed, it has been argued in the Chinese case that it may be used in too inclusive a manner, with what Keane (2007) refers to as the "super-sign" of creativity being applied to areas as diverse as hairdressing, theme parks, and furniture manufacture in order to bolster claims that creative industries are central to a "new China" that can move beyond being the global center of low-cost manufacturing.

Although policy discourses are tending to settle around creative industries, if not necessarily converging, the gulf between creative industries theory and policy discourse and the positions of critical theorists remains wide, and has almost certainly widened over the 2000s. Various arguments have been made that creative industries discourse subtly endorses neoliberal marketisation (Hesmondhalgh 2008), has been inappropriately used as consultancy-speak (Miller 2009), gives a positive spin to the activities and products of global media corporations (Kellner 2009), and ignores the plight of precarious labor (Rossiter 2006). At one level, such critiques can be understood as an extension of debates in the 1990s about cultural policy studies, and whether there is a need to incorporate policy considerations into the study of culture as a field shaped by governmental practices and discourses (Bennett 1998). A lot of the critical debate about creative industries can be seen as a variant of earlier debates about whether a focus upon the pragmatic, ad hoc, ameliorative and "ideas-thick" realm of public policymaking appears inadequate and compromised in the eyes of those who champion the transformative, heroic, programmatic and "ideas-rich" realms of cultural critique and the critical humanities (Cunningham 1992).

However, the stakes have been upped in creative industries debates from the cultural policy debate of the 1990s. One major reason is that larger claims are being made about the broader socioeconomic significance of culture and creativity arising from transformations in the technological and economic substructure of twenty-first-century global capitalism. As a result, creative industries discourse seeks to engage not only with the public sector and regulated cultural industries, but also with a wider range of knowledge and service industries, professions, and practices. This means that the focus has shifted toward whether creative industries are loci of innovation and employment growth in increasingly knowledge-based economies; cultural policy is moving from arts subsidy and advocacy to the center stage of economic growth policies in postindustrial economies, at the level of cities, regions, or nations.

CREATIVE INDUSTRIES AND THE GREAT NEOLIBERALISM DEBATE

At the core of the critical theorists' dissent with creative industries is the claim that it promotes neoliberalism as a political ideology, and that this furthers the hegemony of multinational corporate capital over the cultural sphere. Miller finds that "neoliberal creative industries discourse" has been promoted by "carpet-bagging consultants" pushing a "cybertarian mythology," while "the cultural industries remain under the control of media conglomerates" (Miller 2009, 188, 190, 194). Des Friedman views the rise of creative industries discourse in the United Kingdom as part of a larger project of "the neo-liberalization of media policy," which "is designed to transform the existing balance of power . . . to assist the expansion of private accumulation and to undermine the legitimacy and existence of non-profit and public service media provision" (Freedman 2008, 224).

The claim of complicity with neoliberalism is a more serious charge directed at creative industries arguments than those concerning what industries are included or excluded, or whether it is helpful to differentiate creativity and culture in understanding sectoral dynamics and their policy implications. Neoliberalism as a concept emerged in the late 1990s and early 2000s as an omnibus term used by activists to critique and protest institutions and forums associated with globalization, such as the World Trade Organization and its inaugural meeting in Seattle in 1999, and summits of world leaders such as the Group of Eight (G8) Summit in Genoa in 2001 and the Group of Twenty (G20) Summit in London in 2009. As the critique of neoliberalism developed in international economics in the 1990s, it also referred to what has also been termed the "Washington Consensus," where the application of a common set of policies based around fiscal austerity, privatization of publicly owned assets, and market liberalization was the common recommendation of U.S.-based international institutions such as the International Monetary Fund and the World Bank for developing countries facing economic difficulties (Stiglitz 2002).

Critics would see neoliberalism and globalization as being connected. Scholte (2005, 1) refers to neoliberal globalization as "an economically driven process that should proceed on first principles of private property and uninhibited market forces," and where "other economic rules and institutions are 'political interferences' that undermine market efficiency and should therefore be reduced to a minimum." Neo-Marxists such as David Harvey (2005) identified neoliberalism as a global strategy to reassert the class power of business and economic elites that has its origins in the ideas and policies that shaped the Thatcher and Major governments in Britain from 1979 to 1997 and the Reagan administration in the United States in the 1980s. Harvey (ibid., 2) defined neoliberalism as "[a] theory of political economic practices that proposes that human well being can best be advanced by liberating individual entrepreneurial freedoms and skills within an institutional framework characterized by strong private property rights, free markets, and free trade." Harvey's neo-Marxist critique sits alongside neo-Foucauldian accounts such as that of the British political theorist Nikolas Rose, who identifies neoliberalism with the association (found on the libertarian left as well as among many conservatives) of "notions of freedom, and the associated celebration of the powers of the individual ... whether as discerning customer, enterprising individual, subject of right or autonomous fellow human" (Rose 1999, 64). While this may sound like a somewhat academic debate in some circles, the range of anti-corporate and anti-globalization protests over the 2000s has kept the term on the agenda, while the global financial crisis that began in October 2008 saw such ostensibly establishment figures as the Australian Prime Minister, Kevin Rudd (2009), attributing the severity of the crisis to the impact of neoliberal policies and "extreme capitalism."

Debates about neoliberalism and creative industries are connected through three factors. The first is the "New Labour" administrations that governed Britain under Tony Blair and Gordon Brown. As the Blair Labour government provided much of the early thought leadership about creative industries being in the vanguard of a postindustrial "Cool Britannia," creative industries has tended to be seen as prototypical of that government's "Third Way" ideology. The second factor was the focus upon markets, entrepreneurship, and intellectual property found in the creative industries literature, which, it was argued, did not adequately consider the ways in which capitalist markets could produce inequalities of access or shape cultural forms in ways that may be deemed at odds with wider notions of the public or social good (Hesmondhalgh 2007). Finally, the focus upon new industries, emergent markets, and the small-to-medium enterprise (SME) sector cut across traditional policy divides, where neoliberals focus on the economic case for greater competition and on the opportunities presented by new technologies and reduced public-sector intervention, and the left focuses on the social case for public ownership, regulation, and public subsidy of the arts (Flew 2005b, 2006). Since creative industries did not speak the language of the traditional

left—a mix of cultural Marxism, suspicion of markets and commercial enterprise, and enthusiasm for the regulatory state—it was easy to see it as a feint toward the political dark side among those associating their own positions with those of with the traditional left, wondering whether "neoliberal emphases on creativity have succeeded old-school cultural patrimony" (Miller 2009, 187).

ADDRESSING THE CRITIQUE OF CREATIVE INDUSTRIES AS NEOLIBERALISM

A wider critique of neoliberalism as an explanatory concept is beyond the scope of this article, but points that have been raised in relation to creative industries can be addressed. The association of the creative industries concept with "New Labour" governments in Britain has enabled critics to tie the concept to a wider meta-narrative about neoliberalism as a political-ideological project of dominant economic elites, with terms such as the "new economy" and "creative industries" being viewed as ideological obfuscations designed to disguise the extent to which they had essentially accepted the policies of their conservative predecessors. Both Hesmondhalgh (2007) and Freedman (2008) structure their analysis of media policy since 1980 in precisely these terms, and it is the dominant approach taken by political economists more generally (see, e.g., Curran 2006; McChesney 2008).

The notion that we have been in an era of "neoliberal globalization" or "neoliberal capitalism" since the 1980s has become something of an intellectual truism, established as a given intellectual proposition by virtue of frequent enunciation. Andrew Kipnis (2007) has observed that the number of articles in leading anthropological journals using the term "neoliberal" increased from less than 10 percent in the decade prior to 2002 to 35 percent of articles published between 2002 and 2005. Nonini (2008, 149) notes that:

> The term "neoliberal" has recently appeared so frequently, and been applied with such abandon, that it risks being used to refer to almost any political, economic, social or cultural process associated with contemporary capitalism. . . . A term with so many meanings obviously has great utility, because most progressive scholars can agree that whatever neoliberalism is, they don't like it, and the ambiguity of the term allows discursive coalitions of the like-minded to form without the troublesome bother of having to clarify exactly what it is they oppose or are critical of.

The limits of neoliberalism as a general explanatory framework for global capitalism since the 1980s are evident in the case of China. Contrary to Harvey's (2005) account of Chinese developments since 1980 as "neoliberalism with Chinese characteristics," Nonini (2008) argues that the depth of official commitment to private property rights, free markets, and free trade—to take three baseline commitments of neoliberalism—is limited, contingent, and reversible, particularly if enhancement of any of these were to challenge the power of the Chinese Communist party-state. Moreover, he argues that popular support for a neoliberal policy program in China is virtually nonexistent, reflecting the historically weak position of liberalism as a political philosophy in Chinese society, and that while there may be some support for a "weak" variant of neoliberalism based around support for markets, entrepreneurship, and consumerist values, "the strong version of neoliberalism does not exist in China as a hegemonic project" (ibid., 168). China may be an exceptional case, given its size, its rapid growth rates since 1978, and its significance in the global economy. Moreover, insofar as the Chinese case has parallels, it is in the generally strong support found for the "developmental state" in Asian capitalist systems (Weiss 2003).

Rather than debating the Chinese case at length, the point in raising it is to draw attention to the extent to which universalizing claims about neoliberalism may in fact rest upon a kind of Marxist functionalism, whereby an all-encompassing dominant ideology is developed to "serve" capital in its latest phase, which is deemed to be global and flexible. Nonini (2008, 151) proposes downsizing our claims about neoliberalism and giving them historical, geographical, and cultural specificity. Otherwise, the real risk exists of "assuming that flexible capitalism brings about the very political conditions within nation-states of deregulation and privatisation etc., which it needs for maximum capital accumulation, and . . . that flexible 'capital' has a universal global capacity to do so, and that to do so is somehow 'neo-liberal' governance, restructuring, domination etc., wherever it occurs in the world." We can note here Will Hutton's argument that China requires an "Enlightenment infrastructure" in order to properly develop capitalism, and that "China will only be able to truly compete with the West if it becomes more like us." The response of economist Meghnad Desai to the claim that there is one "true" capitalism based on individualism, liberty, and pluralism that China needs to adopt, is that "Capitalism . . . has accommodated a variety of institutional arrangements and only in the most recent phase of globalization have we thought that an Anglo-Saxon style liberal democracy is its *sine qua non*" (Hutton and Desai 2007). In light of the wide range of work that has sought to critically evaluate different national capitalisms and their responses to globalization (see, e.g., Perraton and Clift 2004), we would share Kipnis's (2007, 387) observation that "to naïvely draw upon all types of analyses of neo-liberalism without noting their contradictions leads to a hodgepodge sort of analysis in which the world as a whole and everything in it appears to belong to a single theoretical category."[1]

CREATIVE INDUSTRIES BEYOND CARICATURES

If we move beyond the crude claim that invoking Tony Blair or "New Labour" proves in itself the neoliberal provenance of creative industries, consideration needs to be given to how evidence of neoliberal strategies manifest themselves in various policies and forms of policy discourse. There is certainly considerable talk about markets, entrepreneurship, competition, and innovation in creative industries policies, but it has been noted that this was not a new thing. Cultural policy since the 1970s had been moving from a supply-side, artist-centered approach to one that gave stronger consideration to consumer demand and cultural markets. One distinctive feature of creative industries as a policy discourse—if not necessarily policies as actually applied in the arts and media sectors—has been increased attention given to the nature of small-to-medium enterprises (SMEs) in the creative industries. Although arts policy has often been oriented toward flagship cultural institutions and major events, and political economists have focused upon the largest commercial and public sector media conglomerates—what Nick Couldry (2006) refers to as "the myth of the mediated centre"—the creative industries have come to evolve what has been termed an "hourglass" structure, with a small number of major players in each sector sitting alongside a myriad of individual enterprises, small companies, and networks of creative talent (Deuze 2006). As these individuals and small groups are relatively new and not highly concentrated, and as "portfolio careers" characterized by multiple jobs across different sectors are often the norm for these segments of the creative workforce, they lack the political power and lobbying clout of big corporations, established trade unions, and traditional arts organizations. Yet there is growing evidence that such loosely configured creative networks are a core source of innovation in the arts, media, and culture, and the challenge has been raised of how policy frameworks can best support such networks that differs from the traditional large-scale institutional domains of media and cultural policies and politics.

Another distinctive issue has been how the Internet and digital media production and distribution models are changing the producer-consumer dichotomy that has long characterized mass communication models and critical theories of the mass media. While cultural studies theorists and critical political economists have long debated the capacity for autonomous agency among media consumers using the products distributed by mass media corporations, the rise of what Yochai Benkler (2006) terms *social production* models based around collaborative networks and peer production are generating new sources of competition, conflict, and contradiction at the heart of industries connected to information, knowledge, communication, culture, and creativity, which are themselves moving to the center of twenty-first-century economies. Benkler (ibid., 122) observes that "social production in general and peer production in particular present new sources of competition to incumbents that produce information goods for which there are now socially produced substitutes." This requires a different approach to the politics of media reform, focused not only on the regulation of media corporations and provision of support for public service media, but also on new flashpoint issues such as the future of copyright and intellectual property law, open-source versus proprietary software, user-generated media content, and questions of open access to repositories of creative content. As Henry Jenkins observes, in what he refers to as *convergence culture*, "the potentials of a more participatory media culture are worth fighting for... [as] convergence culture is throwing media into flux, expanding the opportunities for grassroots groups to speak back to the mass media" (Jenkins 2006, 248). While such concerns have been caricatured as claims that creativity and technology will in and of themselves trump corporate power structures (e.g., Miller 2009), there seem to be very important and current political questions arising out of such power shifts between users and distributors of digital cultural content.

Finally, there is the role of public-sector cultural institutions in the creative economy. The implications of creative industries theories and policy discourses are not necessarily neoliberal ones that "cement the victory of private over public interests" or "limit the scope of public service broadcasting and reduce it to a ghettoized corner of the broadcast market where private operators have no desire (and no compulsion) to go" (Freedman 2008, 224). By pointing to a positive correlation between the development of culture and the creative industries and economic growth and innovation, rather than seeing cultural provision as a rent extracted from the "real" or "productive" economy on the basis of social or cultural value rationales alone, we can begin to think about such cultural institutions as *public-sector social innovation* incubators (Cunningham 2009a). Governments can use their more direct leverage over these institutions to establish them as leading-edge sites for innovations around user-generated content, open access, and a more participatory media culture, and they can operate as a fulcrum for wider changes in the cultural sphere. This is not to say that the influence of governments will invariably be a benign one—the long and debilitating "culture wars" in countries such as the United States and Australia indicate the extent to which such cultural institutions can be hobbled for perceived political gain—or that such initiatives will not be resisted from within the organizational culture of such large, well-established cultural institutions. But it is to say that creative industries theories and policy dis-

courses are not defined by an ideological preference for large commercial institutions over those of the public sector. Rather, they can act as an advocate for the contribution of SMEs, and for the formative role of public-sector cultural institutions as cultural questions move to the fore of globalized knowledge-based economies and societies.

NOTE

1. Such a problem pervades Freedman's (2008, 223) account of the politics of media policy in Britain under the Blair government and the United States during the Bush administration. Despite the author's concerns about "the tendency to treat neo-liberalism as an undifferentiated 'bogeyman' of contemporary capitalism," it can be argued that his account of a diverse range of media policies in the two countries as variants of neoliberalism becomes precisely this. This is most marked in his analysis of public broadcasting, where the overt politicization and defunding of the Public Broadcasting Service (PBS) by the Bush administration in the United States is seen as being essentially similar to the Blair government's promotion of new market opportunities for the British Broadcasting Corporation (BBC). In this and other cases in the book, such as content regulation and censorship, quite different policies are approached as evidence of "the emergence of varieties of neo-liberalism... [where] states are experimenting with and internalizing different aspects of the neo-liberal agenda, contributing to the emergence of 'diversity within convergence.'" In other words, even when policies would appear to be quite different, they are in fact quite the same, all explicable under the rubric of variants of neoliberalism! For a critical review of Freedman, see Flew (2009).

REFERENCES

Americans for the Arts. 2008. *Research services: Creative industries.* http://www.artsusa.org/information_services/research/services/creative_industries/default.asp (accessed August 13, 2009).

Barrowclough, D., and Z. Kozul-Wright. 2008. *Creative industries and developing countries: Voice, choice and economic growth.* London: Routledge.

Benkler, Y. 2006. *The wealth of networks: How social production transforms markets and freedom.* New Haven, CT: Yale University Press.

Bennett, T. 1998. *Culture: A reformer's science.* Sydney: Allen & Unwin.

Bilton, C., and R. Leary. 2002. What managers can do for creativity: Brokering creativity in the creative industries. *International Journal of Cultural Policy* 8:49–64.

Caves, R. 2000. *Creative industries: Between art and commerce.* Cambridge, MA: Harvard University Press.

Couldry, N. 2006. Transvaluing media studies: Or, beyond the myth of the mediating centre. In *Media and cultural theory*, ed. J. Curran and D. Morley, 177–94. London: Routledge.

Coyle, D. 1998. *The weightless world.* Cambridge, MA: MIT Press.

Craik, J. 2007. *Re-visioning arts and cultural policy: Current impasses and future directions.* Canberra: Australian National University e-Press.

Cunningham, S. 1992. *Framing culture: Criticism and policy in Australia*. Sydney: Allen & Unwin.

———. 2002. From cultural to creative industries: Theory, industry and policy implications. *Media International Australia* 102:57–67.

———. 2007. Creative industries as policy and discourse outside the United Kingdom. *Global Media and Communication* 3:347–52.

———. 2008. Creative industries as a globally contestable policy field. *Chinese Journal of Communication* 2:13–24.

Cunningham, S., and P. Higgs. 2008. Creative industries mapping: Where have we come from and where are we going? *Creative Industries Journal* 1:7–30.

Cunningham, S., J. Banks, and J. Potts. 2008. Cultural economy: The shape of the field. In *Cultural economy*, ed. Herman Anheier and Yudhishthir Raj Isar, 15–26. London: Sage.

———. 2009a. Reinventing television: The work of the 'innovation' unit. In *TV studies after TV*, ed. Graeme Turner and Jinna Tay, 83–92. London: Routledge.

———. 2009b. Trojan horse or Rorschach blot? Creative Industries discourse around the world. *International Journal of Cultural Policy* 15:375–86.

Curran, J. 2006. Media and cultural theory in the age of market liberalism. In *Media and cultural theory*, ed. James Curran and David Morley, 129–48. London: Routledge.

David, P., and D. Foray. 2002. An introduction to the economy of the knowledge society. *International Social Science Journal* 171:9–23.

Department of Culture, Media and Sport. 1998. *Creative industries mapping document*. London: DCMS. http://www.culture.gov.uk/creative/creative_industries.html (accessed May 5, 2001).

Deuze, M. 2006. *Media work*. Cambridge, MA: Polity Press.

Donald, J. 2004. What's new: A letter to Terry Flew. *Continuum: Journal of Media and Cultural Studies* 18:235–46.

Edgar, A. 2008. Culture industries. In *Cultural theory: The key concepts*, ed. Andrew Edgar and Peter Sedgwick, 83–84. London: Routledge.

Flew, T. 2002. Beyond *ad hocery*: Defining the creative industries. *Proceedings Cultural Sites, Cultural Theory, Cultural Policy, Second International Conference of Cultural Policy Research*, Wellington, New Zealand, January 23–26, ed. Michael Volkerling, 181–191.

———. 2005a. Creative economy. In *Creative industries*, ed. John Hartley, 344–360. Oxford, U.K.: Blackwell.

———. 2005b. Sovereignty and software: Rethinking cultural policy in a global creative economy. *International Journal of Cultural Policy* 11: 243–60.

———. 2006. The social contract and beyond in broadcast media policy. *Television and New Media* 7:282–305.

———. 2009. A game of two halves [Review of Des Freedman, *The Politics of Media Policy*]. *Australian Journalism Review* 30:125–27.

Freedman, D. 2008. *The politics of media policy*. Cambridge. Polity.

Galloway, S., and S. Dunlop. 2007. A critique of definitions of the cultural and creative industries in public policy. *International Journal of Cultural Policy* 13:17–31.

Garnham, N. 1987. Concepts of culture: Public policy and the cultural industries. *Cultural Studies* 1:23–37.

———. 2005. From cultural to creative industries: An analysis of the implications of the "creative industries" approach to the arts and media policy making in the United Kingdom. *International Journal of Cultural Policy* 11(1):15–29.

Hartley, J. 2005. Introduction. In *Creative industries*, ed. John Hartley, 1–39. Oxford, UK: Blackwell.

Hassan, R. 2008. *The information society*. Cambridge: Polity.

Hesmondhalgh, D. 2007. *The cultural industries*, 2nd ed. London: Sage.

———. 2008. Cultural and creative industries. In *The Sage handbook of cultural analysis*, ed. Tony Bennett and John Frow, 552–69. London: Sage.

Hesmondhalgh, D, and A. Pratt. 2005. Cultural industries and cultural policy. *International Journal of Cultural Policy* 11(1):1–13.

Higgs, P., S. Cunningham, and H. Bakhshi. 2008. *Beyond the creative industries: Mapping the creative economy in the United Kingdom*. London: NESTA.

Holden, J. 2007. *Publicly-funded culture and the creative industries*. London: DEMOS.

Hutton, W., and M. Desai. 2007. Does the future really belong to China? *Prospect Magazine* 130 (January). http://www.prospect-magazine.co.uk/article_details.php?id = 8174 (accessed 8 April 2009).

Jenkins, H. 2006. *Convergence culture: Where old and new media collide*. New York: New York University Press.

KEA European Affairs. 2006. *The economy of culture in Europe*. Study prepared for the Directorate-General for Education and Culture, European Commission. Brussels: KEA European Affairs. http://www.keanet.eu/ecoculturepage.html (accessed April 13, 2009).

Keane, M. 2007. *Created in China: The great new leap forward*. London: Routledge.

Kellner, D. 2009. Media industries, political economy, and media/cultural studies: an articulation. In *Media industries: History, theory and method*, ed. J. Holt and A. Perren, 95–107. Malden, MA: Blackwell.

Kipnis, A. 2007. Neo-liberalism reified: *Suzhi* discourse and tropes of neo-liberalism in the People's Republic of China. *Journal of the Royal Anthropological Institute* 13:383–400.

Kong, L., C. Gibson, L.-M. Khoo, and A.-M. Semple. 2006. Knowledge of the creative economy: Towards a relational geography of diffusion and adaptation in Asia. *Asia Pacific Viewpoint* 47:173–94.

McChesney, R. 2008. *The political economy of media: Enduring issues, emerging dilemmas*. New York: Monthly Review.

McQuire, S. 2001. When is art IT? In *The fibreculture reader: Politics of a digital present*, ed. H. Brown, G. Lovink, H. Merrick, N. Rossiter, D. Teh, and M. Willson, 205–12. Melbourne: Fibreculture Publications.

Miller, T. 2002. A view from a fossil: The new economy, creativity and consumption—Two or three things I don't believe in. *International Journal of Cultural Studies* 7:55–65.

———. 2009. Can natural Luddites make things explode or travel faster? The new humanities, cultural policy studies, and creative industries. In *Media Industries: history, theory and method*, ed. J. Holt and A. Perren, 184–98. Malden, MA: Blackwell.

Mitchell, W., A. Inouye, and M. Blumenthal. 2003. *Beyond productivity: Information technology, innovation, and creativity*. Washington, DC: National Academies.

Nonini, D. 2008. Is China becoming neo-liberal? *Critique of Anthropology* 28:145–76.

O'Connor, J. 2007. *The cultural and creative industries: A review of the literature* (Report for Creative Partnerships). London: Arts Council England.

Perraton, J, and B. Clift. 2004. So Where are national capitalisms now? In *Where are national capitalisms now?* ed. J. Perraton and B. Clift, 195–260. Basingstoke, UK: Palgrave Macmillan.

Pratt, A. 2005. Cultural industries and public policy. *International Journal of Cultural Policy* 11(1):31–44.

———. 2009. The creative and cultural economy and the recession. *Geoforum* 40:495–96.

Rifkin, J. 2000. *The age of access*. London: Penguin.

Roodhouse, S. 2001. Have the cultural industries a role to play in regional regeneration and a nation's wealth? In *Proceedings of AIMAC 2001: 6th International Conference on Arts and Cultural Management*, ed. J. Radbourne, 457–66. Brisbane: Faculty of Business, Queensland University of Technology.

Rose, N. 1999. *Powers of freedom: Reframing political thought*. Cambridge: Cambridge University Press.

Ross, A. 2007. Nice work if you can get it: The mercurial career of creative industries policy. In *My creativity reader: A critique of creative industries*, ed. G. Lovink and N. Rossiter, 17–39. Amsterdam: Institute of Network Cultures.

Rossiter, N. 2006. *Organized networks: Media theory, creative labour, new institutions*. Amsterdam: Institute of Network Cultures.

Rudd, K. 2009. The global financial crisis. *The Monthly* 42:20–29.

Scholte, J. A. 2005. *The sources of neo-liberal globalization* Overarching Concerns Paper No. 8. United Nations Research Institute for Social Development, Geneva, Switzerland.

Siwek, S. 2006. *Copyright industries in the U.S. economy*. Washington, DC: International Intellectual Property Alliance.

Throsby, D. 2001. *Economics and culture*. Cambridge: Cambridge University Press.

———. 2008. Modelling the cultural industries. *International Journal of Cultural Policy* 14:217–32.

Towse, R. 2003. Cultural industries. In *A handbook of cultural economics*, ed. R. Towse, 170–76. Cheltenham, UK: Edward Elgar.

United Nations Committee on Trade, Aid and Development. 2008. *Creative economy report 2008*. Geneva: UNCTAD.

United Nations Educational, Scientific, and Cultural Organisation. 2007. *The 2009 UNESCO framework for cultural statistics (Draft)*. Montreal: UNESCO Institute for Statistics.

Weiss, L. 2003. Bringing domestic institutions back in. In *States in the global economy: Bringing domestic institutions back in*, ed. L. Weiss, 1–37. Cambridge, UK: Cambridge University Press.

Wyszominski, M. J. 2008. The local creative economy in the United States. In *Cultural economy*, ed. H. Anheier and Y. Raj Isar, 199–212. London: Sage.

Index

T - #0715 - 101024 - C0 - 276/216/5 - PB - 9781138841772 - Gloss Lamination